M. Rosseneu, K. Widhalm, and
J. Jarausch (eds.)

Apolipoproteins in Lipid Disorders

Risk Assessment and Monitoring

Springer-Verlag Wien New York

Dr. Maryvonne Rosseneu, Brugge
Prof. Dr. Kurt Widhalm, Wien
Dr. Jochen Jarausch, Mannheim

Kandlgasse 19-21, 1070 Wien
Printed on acid-free paper

With 28 Figures

ISBN-13:978-3-211-82273-9 e-ISBN-13:978-3-7091-9148-4
DOI: 10.1007/978-3-7091-9148-4

Preface

There is increasing evidence for the clinicial value of the apolipoprotein measurements. Besides cholesterol in plasma and lipoprotein fractions, which is currently used as an indicator of cardiovascular risk, the measurement of the AI and B apolipoproteins can provide additional information about the patients' clinical status.

Several studies show that apo B is higher and apo AI is lower in patients with angiographically documented coronary heart disease than in symptomatic patients without coronary heart disease. Moreover, discriminant analysis indicated that the concentration of Apo AI and B in plasma are better discriminators than lipoprotein cholesterol for identifying patients with coronary heart disease. In some studies the apo B/apo AI ratio appears to be a more powerful predictor than individual lipoproteins. In a recent study carried out in men, apolipoproteins AI and B were better correlated with the severity of cardiovascular disease than HDL and LDL cholesterol. The predictive power of apolipoproteins could however not be demonstrated in all studies and the value of apolipoprotein measurements in the field of clinical chemistry is still controversial. This is probably due to discrepancies between the results of various studies, arising from differences in the type of immunoassays, the lack of universal reference materials, differences between study protocols, variations in the selection of patients and in the grading and interpretation of coronary lesions.

The purpose of the papers, first presented at a workshop in Eze (France), September 4, 1990, is to evaluate the clinical significance and benefit of apolipoprotein measurements together with their limitations and the recent technical improvements in the assay procedures.

We are grateful to all contributors to this workshop and also to all those involved in its organization and preparation and would like to extend our sincere thanks to Mrs. M. Grattier from Boehringer Mannheim France and Dr. J. Klüber and Mrs. D. Raab from Boehringer Mannheim Germany for their cooperation and support.

The editors

Contents

Participants – Workshop Apolipoproteins AI and B
Eze (France), September 4, 1990

K. J. Anton, Mannheim, Federal Republic of Germany
K. Astaniou, East Sussex, United Kingdom
E. Casals, Barcelona, Spain
R. Darioli, Lausanne, Switzerland
J. P. Deslypere, Gent, Belgium
H. Drexel, Innsbruck, Austria
G. Eberl, Penzberg, Federal Republic of Germany
P. Fievet, Douai Cédex, France
D. Gnat, Brussels, Belgium
J. A. Gómez Gerique, Barcelona, Spain
M. Grattier, Meylan, France
K. Grüter, Rotkreuz, Switzerland
M. Guyon, Meylan, France
H. Herrmann, Mannheim, Federal Republic of Germany
M. Hubert, Bruxelles, Belgium
F. X. Huchet, Poitiers, France
J. Jarausch, Mannheim, Federal Republic of Germany
J. Karl, Tutzing, Federal Republic of Germany
J. Klüber, Mannheim, Federal Republic of Germany
G. M. Kostner, Graz, Austria
C. LeGood, East Sussex, United Kingdom
P. Lehmann, Mannheim, Federal Republic of Germany
R. Mas, Barcelona, Spain
R. C. Mordasini, Bern, Switzerland
M. Rosseneu, Brugge, Belgium
W. Rupp, Mannheim, Federal Republic of Germany
U. Seifert, Frankfurt, Federal Republic of Germany
K. Widhalm, Vienna, Austria
H. Wieland, Freiburg, Federal Republic of Germany
H. Zawta, Mannheim, Federal Republic of Germany

Abbreviations

ACAT	Acyl-CoA Cholesterol Acyl Transferase
AMI	Acute Myocardial Infarction
apo	Apolipoproteins
BA	Bile Acids
BF	Bottom Fraction
CE	Cholesteryl Ester
CEETP	Cholesteryl Ester Exchange/Transfer Proteins
Ch	Cholesterol
CHD	Coronary Heart Disease
CYM	Chylomicrons
EC	Endothelial Cells
EDRF	Endothelial Derived Relaxing Factor
FC	Free Cholesterol
FFA-Alb	Free Fatty Acid-Albumin-Complex
FH	Familial Hypercholesterolemia
GAG-LDL	Glycosaminoglycan-LDL
HL	Hepatic Lipase
HDL	High Density Lipoproteins
HDL-C	High Density Lipoprotein Cholesterol
HMG-CoA Reductase	β-Hydroxy-β-Methyl-Glutaryl-Coenzym-A-Reductase
HSL	Hormon Sensitive Lipase
HTGL	Hepatic Triglyceridelipase
IDL	Intermediate Density Lipoproteins
LCAT	Lecithin Cholesterol Acyl Transferase
LDL	Low Density Lipoproteins
LDL-C	Low Density Lipoprotein Cholesterol
Lp	Lipoprotein
LPL	Lipoprotein Lipase
MI	Myocardial Infarction
PAF	Platelet Activating Factor
PDGF	Platelet Derived Growth Factor
Pg	Plasminogen

PL	Phospholipids
REM	Remnants
SMC	Smooth Muscle Cells
TC	Total Cholesterol
TG	Triglycerides
TPA	Tissue-type Plasminogen Activator
VLDL	Very Low Density Lipoproteins

Biochemistry and pathophysiology of human plasma apolipoproteins

G. M. Kostner

Institute of Medical Biochemistry, Graz, Austria

Summary

In order to learn more about the impact of lipoprotein metabolism on premature atherosclerosis, analysis not only of plasma lipids but also that of apolipoproteins have been performed in the past. Now we have reached a stage, where functional aspects become more and more prominent. Molecular biology has provided tools to routinely assay structural alterations of the proteins involved in the lipoprotein metabolism. In the near future, the DNA of patients might even be probed for abnormalities in the promotor region of apolipoprotein genes or their regulatory elements. Yet, without doubt, for clinical chemistry, the measurements of apolipoprotein concentrations will remain a key for classifying atherogenic lipoprotein constellations.

Keywords: Apolipoproteins, review, biochemistry, pathophysiology.

Introduction

Plasma apolipoproteins function as transport lipids which are important for the transport of cholesterol and cholesteryl esters (CE), triglycerides (TG), phospholipids (PL) as well as lipid soluble vitamins, vit. A, D, E, K [8, 9, 10, 11]. All these lipids possess important physiological functions, serving as fuel for energy storage and production, as components of cell membranes, precursors for steroid hormones and bile acids and as functional groups in co-enzymes. With the exception of vitamins, all lipids mentioned above may be biosynthetized in animals and men in amounts sufficient to meet the bodies need. There is, however, a great deal of energy necessary to synthesize some of these lipids, notably cholesterol. Nature therefore provided us with biological mechanisms to absorb and utilize dietary triglycerides and cholesterol. Such mechanisms are necessary for many animal species and also for humans living in underdeveloped countries, where dietary supply of energy is scarce. In the western civilized world, however, food rich in fat, cholesterol and energy is abundant and may cause adverse effects – especially in persons with genetic pre-disposition to abnormalities in lipid metabolism and/or to atherosclerosis.

Plasma lipoproteins and apolipoproteins

Lipoproteins may be characterized by their behaviour in the ultracentrifuge (Table 1). Although the kind and amount of individual lipids in a given lipoprotein density class determines the content of apolipoproteins, the metabolism of lipoproteins is triggered by the protein moiety. In addition, there are numerous other proteins, e.g. enzymes, cofactors and receptors involved in the intravascular lipoprotein metabolism. The most important ones are shown in Table 2. Apolipoproteins function as structural proteins to keep lipid particles in solution. Each density class of normal fasting human plasma contains a characteristic apolipoprotein pattern which is shown in Fig. 1. LDL for example, consists of > 90% of apo B, and HDL of about 90% apo AI plus apo AII. Apolipoproteins also function as ligands for cell surface receptors targeting lipoproteins into specific organs. Finally, apolipoproteins activate or regulate enzymes. The enzymes listed in Table 2 are intimately involved in the intracellular lipid and lipoprotein metabolism and play a key role in the structure, composition and plasma concentration of lipoproteins.

Table 1. Main plasma lipoproteins

Lipo-protein	Density	Plasma concentr. (mg/dl)	Main apolipo-protein	Composition (%)			
				Protein	PL	Ch	TG
CYM	<1.000	<10	A, B-48, C, E	2	4	5	89
VLDL (pre-β)	<1.006	50–200	B, C, E	10	16	17	57
IDL	<1.019	5–15	B, C, (E)	16	20	33	31
LDL (β-Lp)	<1.063	200–300	B	25	23	47	5
Lp (a)	<1.100	1–100	B, (a)	35	20	41	4
HDL-2	<1.125	50–150	A	41	35	19	5
HDL-3	<1.210	100–200	A	57	23	16	4

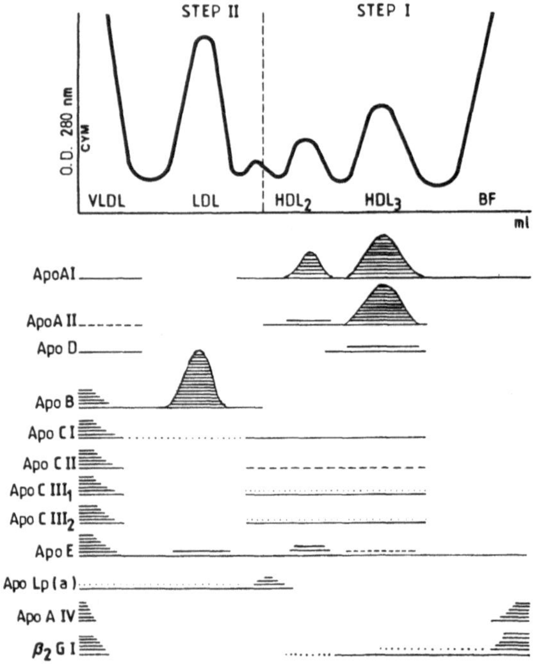

Fig. 1. Apolipoprotein patterns in the density classes of normal human plasma

The cellular proteins listed in Table 2 regulate the uptake as well as further metabolic steps of cholesterol in human tissues, thus playing a key role in atherogenesis.

Lipoprotein metabolism

Lipid absorption

Almost 100% of the ingested amounts of the lipid classes mentioned above are absorbed – except cholesterol. The fraction of cholesterol absorbed by the gastrointestinal tract is approximately only 50%, but interindividual differences occur, ranging from 25–75%. As shown later, the dietary content of cholesterol and saturated fatty acids are the major

Table 2. Proteins involved in the lipoprotein metabolism

Protein	Plasma conc. (mg/dl.)*	Function
1. Apolipoproteins		
Apo A I	120 × 140	Structure protein; activation of LCAT
Apo A II	35–50	unknown
Apo B	80–120	Structure protein; ligand for the B/E-receptor; necessary for lipid absorption and selection
Apo C	5–10	Regulation of lipoprotein lipase activity
Apo D	4–8	Reverse cholesterol transport
Apo E	4–6	Ligand for the B/E-receptor and E-receptor; reverse cholesterol transport
Apo (a)	0–50	Unknown; most atherogenic apolipoprotein
2. Enzymes		
LPL	<1	Hydrolysis of triglycerides in chylomicrons and VLD
HTGL	1–2	Hydrolysis of triglycerides in IDL and HDL-2
LCAT	2–4	Production of cholesteryl esters in plasma
CEETP	1–2	Exchange and transfer of neutral core lipids
3. Cellular proteins		
HMG-CoA reductase		Key enzyme for the cholesterol biosynthesis
B/E-receptor		Binding of apo B and apo E containing lipoproteins
E-receptor/CYM-remnant-R		Binding of apo E containing lipoproteins
Scavenger receptor		Binding of metabolically altered lipoproteins notably to RES-cells
ACAT		Production of intracellular CE

* Average concentration in fasting normolipemic persons

determinants of plasma cholesterol values which mainly explain the variation in morbidity of hypercholesterolemia and atherosclerosis.

Intravascular lipoprotein metabolism

A schematic view of the lipoprotein metabolism is displayed in Fig. 2: triglyceride-rich lipoproteins are secreted from the gut into the lymph in the form of chylomicrons. The latter are attacked by the enzyme lipoprotein lipase (LPL) yielding "core-" and "surface-remnants". The core remnants are rich in CE, TG and apo E, and are immediately catabolized in the liver. The liver itself synthesizes the so-called "endogenous" very low density lipoproteins (VLDL) from components of chylomicrons. VLDL are hydrolyzed by LPL in a similar way as chylomicrons yielding intermediate density lipoproteins (IDL) and finally low density lipoproteins (LDL). LDL have a physiological half life of several days, are very rich in CE and cholesterol, and together with IDL and possibly chylomicrons remnants, the most atherogenic lipoproteins known! A high LPL-activity warrants fast catabolism of the atherogenic remnants and IDLs!

High density lipoproteins (HDL) may be secreted by liver and intestine in a nascent form, representing surface material, but are also formed during lipolysis of chylomicrons and VLDL. There are two HDLs known: HDL-2 and HDL-3. HDL-2 represents the most anti-atherogenic fraction in blood; high LPL activity promotes the formation of HDL-2!

The enzyme lecithin cholesterol acyl transferase (LCAT), as well as the conversion of HDLs into each other, also plays a key role in the production of CE. The CE formed in HDL-3 by LCAT is shuttled to the apo B containing lipoproteins (IDL and LDL) by the action of cholesteryl ester exchange/transfer proteins (CEETP), loading them with core lipids. In addition to taking up free cholesterol from peripheral tissue which is converted to CE by LCAT, HDL may also enter the cells and load themselves with intracellular cholesterol and apo E; these HDLs are then retroendocytosed and targeted to the liver, where cholesterol is used for the synthesis

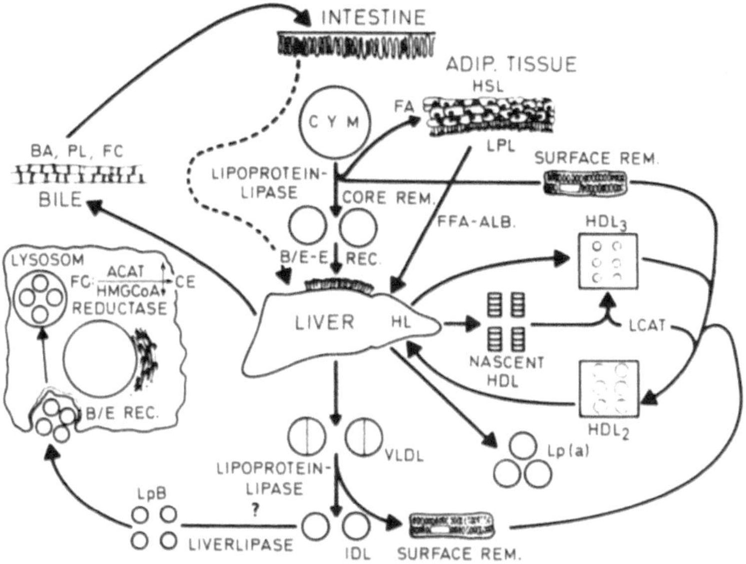

Fig. 2. Schematic view of the lipoprotein metabolism

of bile acids. The whole process is called "Reverse cholesterol transport" and is considered to be the most important "anti-atherosclerotic mechanism".

The role of lipoprotein receptors

Lipoprotein receptors play a central role in the efficient lipid and lipoprotein uptake by cells. Such receptors have been claimed to exist for a variety of apolipoproteins, but there are only three types characterized to an extent to mention them here:

The best known lipoprotein receptor is the apo B/E-receptor detected by Brown and Goldstein [3]. Apo B/E-receptors bind apo B and apo E containing lipoproteins notably IDL, LDL and HDL-E. Although the adrenal cortex is the organ with the highest apo B/E-receptor density, the liver is the most important organ for the catabolism of lipoproteins via specific apo B/E-receptor interaction. As indicated in Fig. 2, LDLs bound to the apo B/E-receptor are taken up by the cells, hydrolyzed in lysosomes and the intracellular free cholesterol suppresses de novo cholesterol synthesis by inactivating HGM-CoA reductase. The intracellular cholesterol is used for the formation of cell membranes as well as for bile acid or sterol hormone synthesis. Excessive free cholesterol may be esterified by LCAT and stored intracellularly in the form of CE. If this latter mechanism prevails, cells stop the expression of apo B/E-receptors on their surface leading to the accumulation of plasma LDL. A high apo B/E-receptor activity warrants fast LDL catabolism and low plasma cholesterol levels!

The second relevant receptor is the apo E receptor or chylomicron-remnant receptor. It is only found on liver cells and is responsible for the fast and efficient uptake of chylomicrons- and VLDL-core remnants and thus the removal of potential atherogenic lipoproteins from the circulation. The apo E-receptor activity is not regulated by extracellular plasma lipids and lipoproteins.

Finally, scavenger receptors play a key role in atherosclerosis! There seems to exist a variety of scavenger receptor on monocytes/macrophages, endothelial cells as well as smooth muscle cells [1]. Scavenger receptors are responsible for the catabolism of all sorts of modified lipoproteins, such as oxidized LDL, malon dialdehyde LDL, GAG-LDL complexes as well as "β-VLDL", which cannot be cleared by the apo B/E-receptor. Although scavenger receptors may be considered as "good guys", they promote the accumulation of CE and foam cell formation in the arterial wall if the system is saturated. High plasma concentrations of cholesterol are a consequence of slow LDL removal; this leads to modification of LDL and activation of the scavenger pathway! High plasma HDL partially counteracts this process by depleting cells from cholesterol.

Pathophysiology of lipoprotein metabolism

There are numerous ways by which the physiological steps listed above may be deranged leading to cholesterol deposition and atherosclerosis. The most important genetic defects are shown in Table 3.

The best studied defects are those affecting plasma apo B levels: *Familial hypercholesterolemia* (FH = Type II A hyperlipoproteinemia) is characterized by the lack of the apo B/E-receptor [3]. Heterozygotes affected by a single gene defect express only half the amount of apo B/E-receptor. As a consequence, plasma TG and LDL-C values are twice as high as normal. Homozygotes have 3–5 times normal values and are prone to atherosclerosis and cardiac death at the age of < 20 years.

Apo B 3500 variant: Measuring the apo B/E-receptor activity in cultured skin fibroblasts with endogenous or exogenous LDL, values spanning a wide range are found. This reflects variations in the affinity of the apo B/E-receptor for apo B, caused by structural aberrations of the receptor or the ligand. Apo B is a large molecule. Thus, a number of genetic variants may be anticipated. In fact, Innerarity et al. [6] have recently described in apo B variant with a point mutation in the amino acid 3500, which does not bind to the receptor. The phenotype is indistinguishable from classical Type II hyperlipoproteinemia, the LDL-receptor defect. From theoretical considerations, one may anticipate that more of such variants exist, with gradual differences in LDL binding.

The second role of apo B, the secretion of TG/CE-rich lipoproteins from the liver and the intestine becomes apparent in abetalipoproteinemia, where no lipoproteins of d < 1.063 are found. On the other hand, in Andersons disease the processing of apo B-100 to apo B-48 seems to be impaired and only chylomicron synthesis is lacking.

Today, we know up to 10 different genetic variants of apo B with the synthesis of truncated forms. These reach from

Table 3. Diseases related to genetic defects

Disease	Synonym	Genetic defect
1. LPL-deficiency	Type I HLP	Lack of functional lipoprotein lipase
2. Familial hypercholesterolemia	Type IIA HLP	B/E-receptor defect
3. Familial combined hyperlipoproteinemia	Type IIB HLP	Over-production of TG and cholesterol
4. Familial dysbetalipoproteinemia	Type III HLP	Structural defect of apo E plus over-production of VLDL
5. Familial hypertriglyceridemia	Type IV HLP	Over-production of VLDL
6. Mixed type	Type V HLP	Polygenic defects
7. Familial hypoalphalipoproteinemia		Unknown
8. Familial hyper-LP (a)		Over-production of Lp (a)

apo B-25 up to apo B-90, which means that only 25% or 90% respectively of the N-terminal portion of apo B-100 is synthesized. From these mutants we learned that at least 30% of the N-terminal sequence is necessary for the LDL assembly.

Apo E variant: The function of apo E is to target chylomicron remnants and other apo E containing lipoproteins into the liver. Utermann [13] was the first to recognize genetic variants of apo E (apo E 2) which cause dyslipoproteinemia with severe hypercholesterolemia and hypertriglyceridemia. These patients are at a high atherosclerosis risk.

Apo E, however, also seems to be involved directly in the absorption of cholesterol. Individuals with apo E-IV absorb strikingly more dietary cholesterol as compared to apo E-III or E-II individuals. The absorption rate ranges from approximately 75% down to 25%. This seems to be one major mechanism which controls plasma cholesterol levels. In addition to common forms, there are less frequent abnormalities described in the literature.

Tangier disease is caused by a defect in the retroendocytosis of HDL from cholesterol loaded cells. These patients exhibit an increased apo A and HDL synthesis rate but a greatly enhanced catabolism. This emphasizes the role of HDL in the scavenger pathway.

Finally, Lp(a) must be considered as one of the most atherogenic lipoproteins. Today we know virtually nothing about the function of Lp(a) (for a review see [12]). However, it is known that Lp(a) is metabolized independently from VLDL or LDL. From the fact that some individuals have Lp(a) values of 100 mg/dl or more corresponding to 30–40 mg/dl cholesterol, it is apparent that Lp(a) might be responsible for > 10% variation of plasma cholesterol levels. In this respect it is interesting that Lp(a) levels cannot be reduced by hypocholesterolemic drugs which increase the number of apo B/E-receptors in the liver.

In Table 3, only the most common forms of lipoprotein abnormalities are listed which are known to be atherogenic. As pointed out above, in addition to the observation that gradual differences in the severity of these abnormalities exist, patients may also be affected by multiple gene defects yielding various kinds of mixed forms.

Secondary hyperlipoproteinemias

Although secondary hyperlipoproteinemias may be triggered by some diseases in genetically unaffected individuals (Table 4), they manifest themselves in an early stage of the respecting disease, if genetic predispositions are present. Examples are the late onset of phenotypic hyperlipoproteinemia Type-III in hypothyroidism, Type-IV and Type-V in diabetes mellitus and gout, as well as severe hypertriglyceridemias in kidney diseases.

Lipid metabolism and atherogenesis

Numerous epidemiological studies have demonstrated the close link of increased plasma cholesterol with cardiac deaths [2, 4, 14] and nobody seems to question anymore that derangements in lipid and lipoprotein metabolism are the most common triggers of atherosclerosis, coronary heart diseases (CHD) and myocardial infarction (MI). Although there exist > 250 risk factors for these diseases [14], most of them directly or indirectly relate to hyper- and dyslipoproteinemias. As the lipoprotein metabolism is triggered by apolipoproteins, it is plausible that any structural change, which in most cases is genetically determined, may have profound effects on cell function and dysfunction. There are four cell types involved in the atherogenesis, which are listed in Table 5.

Apolipoprotein mutants

Today apolipoprotein mutants are known from all apolipoproteins. Point mutations may lead to changes in the electric charge which can be demonstrated in the case of apo AI by isoelectric focusing. Some of the apo AI mutants are connected with a reduced LCAT activation, leading to an impaired reverse cholesterol transport. Apo B mutants have

Table 4. Secondary hyperlipoproteinemias

Trigger	Common hyperlipoproteinemia phenotype
Diabetes mellitus (type II)	Type IV and V; hypo-
Gout	Type IV; hyper-
Kidney diseases	Type III, IV (V)
Liver diseases	LP-X, LCAT deficiency, β-VLDL
Alcohol abuse	Hyper-TG; Type I or V
Hypothyroidism	Type III
Pregnancy	Hypertriglyceridemia and -cholesterolemia

Table 5. Cells involved in atherogenesis

Cell type	Pathological event
Endothelial cells	Desquamation if membranes are rigid; changes in prostacyclin and EDRF production
Smooth muscle cells	Proliferation in response to PDG; formation of foam cells
Platelets	Production of PDGF and thromboxane; clot formation; secretion of histamine, 5-HT and others
Leucocytes	Monocytes/macrophages are loaded with cholesterol and form foam cells; secretion of chemotactic substances (leukotriens); promote peroxidation of lipids

been known for many years and have originally been detected by xenoantibodies. Genetic variants of apo B were previously characterized by "Ag-factors". Because of the huge molecular weight of apo B in most cases, point mutations have little effect on the metabolism, one exception being the B-3500 mutant [6]. On the other hand, if the apo B-100 protein is not fully transcribed due to frame shifts or insertion of stop codons into the apo B gene, the truncated forms of apo B may be found in the plasma. Interestingly enough, the presence of such truncated apo B is not linked with hyper-, but with hypobetalipoproteinemias.

Lipoprotein Lp(a)

Lp(a) was recognized as a risk factor for atherosclerotic diseases about 20 years ago [12]. However, only recently the mysteries of this lipoprotein began to unravel when it turned out that Lp(a) may interfere with fibrinolysis in vivo. Lp(a) is very similar to LDL, but has an additional apolipoprotein, apo(a) attached via one disulfide bridge (Fig. 3). In various isoforms of plasminogen (Pg) kringle 4 may occur in 15–40 repeated copies. The isoforms are responsible for the replacement of Pg from fibrin clots, but may also interfere with plasminogen activation by tissue-type plasminogen activator. In addition, Lp(a) has been found to interact strongly with connective tissue material such as glycosaminoglycans, fibronectin and tetranectin. Once Lp(a) is complexed to such substances it is taken up by macrophages promoting foam cell formation and lipid deposition in arteries.

The physiological function of Lp(a) is unknown, and it appears that individuals without any detectable Lp(a) in plasma have a completely normal lipid metabolism. On the other hand, Lp(a) levels exceeding 25–30 mg/dl are considered to be harmful and individuals with coinciding other atherosclerosis risk factors are prone to early MI and stroke. It is

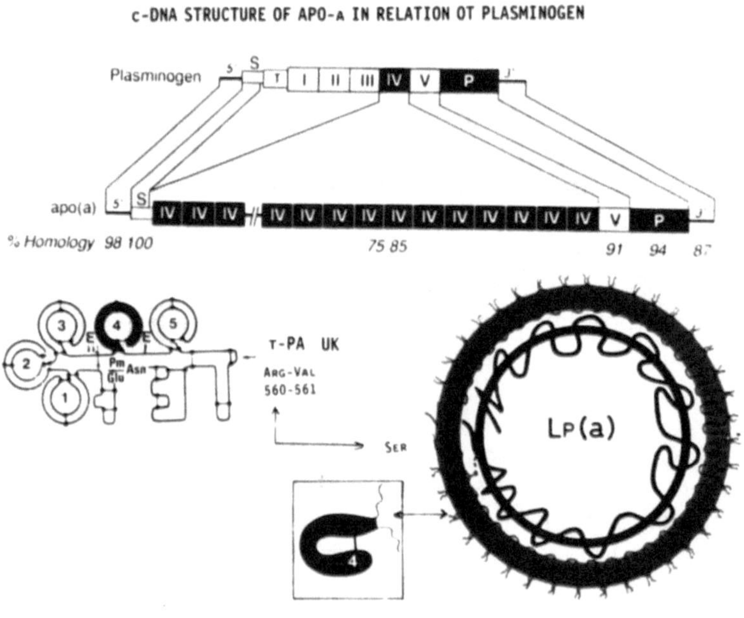

Fig. 3. The structure of Lp(a)

therefore recommended to measure Lp(a) values in addition to other parameters in individuals with familial risk.

Acknowledgements

Work cited in this article has been supported by the Austrian Research Foundation (Project no. S-46) as well as by the Austrian National Bank, project no. 3382.

References

1. Fogelman AM, Haberland ME, Seager J, et al (1981) Factors regulating the activities of the LDL-receptor and the scavenger receptor on human monocyte-macrophages. J Lipid Res 22: 1131–1141
2. Frick MH, Elo O, Haapa K, et al (1987) Helsinki Heart Study: primary prevention trial with gemfibrozil in middle-aged men with dyslipidemia. N Engl J Med 317: 1237–1245
3. Goldstein JL, Brown MS (1982) Lipoprotein receptors: genetic defence against atherosclerosis. Clin Res 30: 417–426
4. Gordon T, Castelli WP, Hjortland MC, et al (1977) High density lipoprotein as a protective factor against coronary heart disease: the Framingham Study. Am J Med 62: 707–714
5. Hopkins PN, Williams RH (1981) A survey of 246 suggested coronary risk factors. Atherosclerosis 40: 1–52
6. Innerarity TL, Weissgerber KH, Arnold KS, et al (1987) Familial defective apolipoprotein B-100: low density lipoproteins with abnormal receptor binding. Proc Natl Acad Sci USA 86: 6919–6923
7. Yound SG, Hubl ST, Smith RS, et al (1990) Familial hypobetalipoproteinemia caused by a mutation in the apolipoprotein B gene that results in a truncated species of apolipoprotein B (B-31). J Clin Invest 85: 933–942
8. Kostner GM (1981) HDL-Cholesterin als Schutzfaktor gegen koronare Herzkrankheit. Internistische Welt 8: 311–320
9. Kostner GM (1984) Fettstoffwechsel und Atherosklerose. Z Allgemeinmedizin 26: 1087–1096
10. Kostner GM (1987) Diagnostik der Fettstoffwechselstörungen. Therapiewoche (Österr Ed) 3: 241–257
11. Kostner GM (1988) Pathobiochemie der Dyslipoproteinemien. Münch Med Wschr 130/14: 251–255
12. Kostner GM (1990) The physiological role of LP(a). In: Scanu AM (ed) Lipoprotein(a). Academic Press, New York, pp 183–204

13. Utermann G (1987) Apolipoprotein E polymorphism in health and disease. Am Heart J 113: 433–440
14. NIH Publication (1984) The lipid research clinics coronary primary prevention trial. JAMA 251: 351–374

Author's address: Dr. G. M. Kostner, Institute of Medical Biochemistry, University of Graz, Harrachgasse 21/III, A-8010 Graz, Austria.

Clinical significance of apolipoprotein measurements

M. Rosseneu

Department of Clinical Chemistry, A.Z. St-Jan, Brugge, Belgium

Summary

Apolipoprotein measurements have been applied to the estimation of coronary risk in patients and also to the characterization of various kinds of dyslipidemia, where apolipoprotein profiles are altered in a significant way. The concentration of apolipoprotein B is increased in various types of primary and secondary dyslipidemia including: Type II, III and V primary dyslipidemia, chronic renal failure and Type I, insulin-dependent diabetes. Apo AI concentrations are decreased in Type I, II B and IV dyslipidemia, as well as in liver disease. Apo E is increased in Type III and V dyslipidemia.

The combination of apolipoproteins and lipids quantitation enables a better discrimination between different kinds of dyslipidemia and the choice of an appropriate dietary or drug treatment.

Keywords: Clinical significance, dyslipidemias, coronary risks, apo AI/B-ratio, cerebrovascular and cardiovascular disease.

Introduction

The abnormal levels of the major lipids: cholesterol, phospholipids and triglycerides observed in many dyslipidemic states are mostly accompanied by defects in the function and distribution of apolipoproteins. The combined quantification of the plasma lipids and apolipoproteins can thus provide a more accurate picture of the disease and enable a better differentiation between the various kinds of dyslipidemias.

From a clinical point of view, apolipoprotein measurements have two main purposes: to estimate coronary risk on one side, and to aid the diagnosis of several primary disorders of lipoprotein metabolism on the other. In this paper we will review the major applications of these assays.

Apolipoproteins and coronary risk

Several case-control studies comparing MI survivors to a group of normal controls, have shown that apo AI levels are decreased and those of apo B are increased in the MI survivors group [3, 8, 10, 14]. In most studies, the discriminative power of apo AI and apo B and especially of the apo AI/B ratio was superior to that of cholesterol and HDL-cholesterol (HDL-C) [3, 10].

In patients with heart disease undergoing coronarography, the plasma apo B concentrations were significantly higher than in a control group without lesions [13, 18]. Apo AI levels were decreased and some studies even showed a positive relationship between apo B increase and severity of the lesions [18].

A further application for the apolipoprotein assays is the monitoring and the early risk detection in off-springs and relatives of patients with coronary heart disease (CHD). We compared the plasma lipids and apolipoproteins in young adults whose father suffered a well-documented MI before 50 years of age, to those of an age and sex-matched control group without any family history of CHD [2, 15]. Lower apo AI levels were the only significant difference between the two groups. They were mainly due to a decrease in the HDL_2 levels in the off-springs of the MI survivors [15]. Similar results were reported by Freedman et al. [4] in a recent study.

A pilot study conducted in newborns [16, 20] enabled the detection of families with decreased apo AI levels and/or increased apo B values both in the parents and children, thus confirming the predictive value of these parameters.

Apolipoproteins in primary dyslipidemias

Hyperlipoproteinemias

Familial dyslipidemias were originally classified by Fredrickson on the basis of the lipoprotein electrophoretic migration patterns [5]. Since then, biochemical defects underlying most of the classical forms of dyslipidemia have been elucidated [7]. Most dyslipidemias are accompanied by changes in the plasma levels of the apolipoproteins, as summarized in Fig. 1. Type II dyslipidemia is characterized by elevated cholesterol and apo B levels in plasma. Triglycerides are either normal in type II A or elevated (II B). Brown and Goldstein [6] have shown that defects in the lipoprotein B/E-receptor activity are responsible for the abnormal patterns in these patients.

Another form of dyslipidemia, called hyperapobetalipoproteinemia, was reported by Sniderman et al. [19]. As a consequence of an over-production of apo B, compared to LDL-cholesterol (LDL-C), these patients have small, dense, apo B-rich LDL particles, which are associated with an increased risk for CHD.

A single mutation in the apo E sequence, if combined with another unknown risk factor, can lead to the appearance of a type III dyslipidemia. This disease is characterized by high apo E and B levels due to the accumulation of VLDL remnants in plasma as a consequence of the decreased affinity for the apo B/E receptor [1].

An impaired conversion of VLDL to LDL and an oversynthesis of triglyceride-rich particles probably accounts for the accumulation of chylomicrons and VLDL in type V and that of VLDL in type IV dyslipidemias. A defective lipase activity and an apo CII deficiency lead to the appearance of type I dyslipidemia, characterized by high triglyceride levels.

The major advantages of the apolipoprotein assays for the characterization and diagnosis of the various forms of dyslipidemia, is to provide more sensitive and detailed information concerning the composition and distribution of the

Primary hyperlipidemias						
Phenotype	I	II_A	II_B	III	IV	V
Lipoprotein abnormality	CM ↑	LDL ↑↑	LDL ↑↑ VLDL ↑	IDL ↑↑	VLDL ↑	CM ↑ VLDL ↑
Biochemical defect	Apo CII or LPL deficiency	LDL-receptor defect	LDL-receptor defect	Apo $E_{2/2}$ phenotype + ?	?	?
Apoprotein						
Apo AI	↓↓	N	N	N	↓	↓
Apo AII	↓↓	N	N	(↓)	(↓)	(↓)
Apo B	↓	↑↑	↑↑	↑ (↑)	↑	↑ (↑)
Apo CII	↑	N	↑	↑	↑	↑↑
Apo CIII	↑	N	↑	↑	↑	↑↑
Apo E	↑	(↑)	↑	↑↑	↑	↑↑

Fig. 1. Lipoprotein and apolipoprotein abnormalities in primary dislipidemias. *CM* chylomicrons; *N* normal; *LPL* lipoprotein lipase

abnormal lipoproteins which accumulate in the patient's plasma. Figure 2 illustrates the changes observed in the apo AI, AII, B and E concentrations in the various kinds of dyslipidemias. Apo AI is decreased in types I, II B and IV, where triglyceride-rich particles accumulate. Apo B levels are increased in types II, III and V where cholesterol-rich lipoproteins are elevated. The measurement of the two apolipoproteins and the calculation of the apo AI/B ratio therefore provides a sensitive parameter to monitor modifications in the lipoprotein patterns, compared to a normolipemic state. Apo E levels are elevated in type III and type V dyslipidemias,

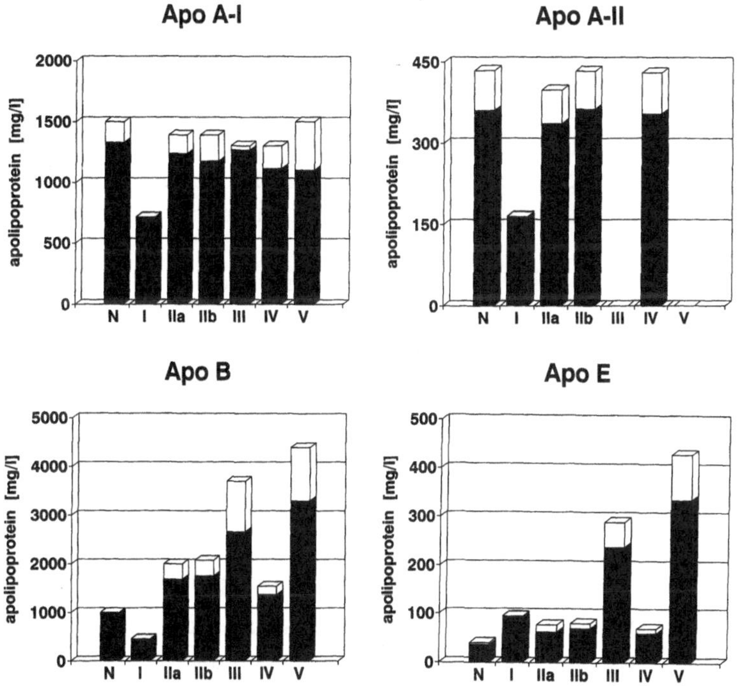

Fig. 2. Plasma concentrations of apolipoproteins AI, AII, B and E in primary dyslipidemias classified according to Fredrickson. Solid bars = mean values; open bars = + SD range

while the apo AII concentrations change only in type IV dyslipidemia.

Hypolipoproteinemias

Abetalipoproteinemia

This primary form of hypolipoproteinemia is characterized by the absence of apo B 100 in plasma [9]. As apo B 48 is also mostly absent, the patients have fat malabsorption problems as well as a deficiency in fat soluble vitamins, leading to neurological problems. The measurement of apo B 100 levels in the proband and in the parents is useful to confirm the case.

Hypobetalipoproteinemia

In this disease, low levels of LDL are present and the immunological assay of apo B is useful to distinguish between low levels and the total absence of the LDL fraction. One or both parents can be affected, whereas they should have normal levels in the case of abetalipoproteinemia.

Hypoalphalipoproteinemia

Familial hypoalphalipoproteinemia was described in families with premature CHD [18]. The probands have low HDL-C and apo AI levels but normal LDL-C and apo B. The origin of the disease is not clear and decreased synthesis as well as increased clearance have been proposed. Mutations of apo AI can also lead to hypoalphalipoproteinemia [17].

Apolipoproteins in secondary dyslipidemias

Aside from the primary dyslipidemias, caused mostly by genetic factors, secondary dyslipidemias can also develop as a consequence of different pathological states, or they can be induced by dietary factors. Abnormal apolipoprotein patterns can appear during renal and liver disease, diabetes mellitus and thyroid dysfunction. These patterns usually normalize when the pathological states are reversed.

Chronic renal failure, treated either by hemodialysis or by continuous ambulatory peritoneal dialysis, is accompanied by an increase in the VLDL and LDL fractions and a decrease of the HDL lipoproteins. The apo B levels are increased and the apo AI concentrations decreased in these patients.

Different types of liver disease also affect the apolipoprotein levels. Acute hepatitis decreases the apo AI and AII levels due to impaired protein synthesis. In chronic hepatitis, the apo AI and AII synthesis is normal, but the hepatic lipase activity is decreased. In liver cirrhosis, the apo AI, AII and B 100 concentrations are decreased due to impaired liver synthesis of these proteins.

Secondary dyslipidemias can also result from diabetes mellitus. In type I, insulin-dependent diabetes, type IV dyslipidemia frequently occurs and is accompanied by an elevation of apo B, CII and CIII concentrations and by decreased apo AI levels. The lipoprotein patterns usually normalize under proper treatment of the disease. In type II, non-insulin dependent diabetes, type V dyslipidemia frequently occurs. In both cases these dyslipidemias increase the risk for CHD in the diabetic patients.

The apolipoprotein levels are also modulated by the thyroid function. Apo B is decreased in hyperthyroidism due to increased catabolism, while the reverse is observed in hypothyroid patients [11, 12].

Conclusion

The plasma apolipoprotein levels are affected in many pathological states which are accompanied by an increased risk of developing atherosclerotic lesions. They therefore represent a class of sensitive parameters for the detection and monitoring of individuals at risk for cerebrovascular and cardiovascular diseases.

References

1. Brown MS, Goldstein JL, Fredrickson DS (1983) Familial type 3 hyperlipoproteinemia (dyslipoproteinemia). In: Stanbury JB, Wyngaarden JB, Fredrickson DS, Goldstein JL, Brown MS (eds) The metabolic basis of inherited disease, 5th edn. McGraw-Hill, New York, 299: 1421–1424
2. De Backer G, Hulstaert F, De Munck K, Rosseneu M, Van Parijs L, Dramaix M (1986) Serum lipids and apoproteins in students whose parents suffered prematurely from a myocardial infarction. Am Heart J 112: 478–484
3. De Backer G, Rosseneu M, Deslypere JP (1982) Discriminative value of lipids and apoproteins in coronary heart disease. Atherosclerosis 42: 197–203
4. Freedman DS, Srinivasan SR, Shear CL, Franklin FA, Webber LS, Berenson GS (1986) The relation of apolipoproteins AI and B in children to parental myocardial infarction. N Engl J Med 315: 721–726
5. Fredrickson DS, Levy RI, Lees RS (1967) Fat transport in lipoproteins: an integrated approach to mechanisms and disorders. N Engl J Med 276: 273–281
6. Goldstein JL, Brown MS (1982) The LDL receptor defect in familial hypercholesterolemia. Implication for pathogenesis and therapy. Med Clin North Am 66: 335–362
7. Havel RJ (1982) Familial dysbetalipoproteinemia. New aspects of pathogenesis and diagnosis. Med Clin North Am 66: 441–454
8. Heiss G, Tyroler HA (1983) Are apolipoproteins useful for evaluating ischemic heart disease? A brief overview of the literature. In: Lippel K (ed) Proceedings of the workshop on apolipoprotein quantification. NIH Publication No. 83-1266, Bethesda, pp 7–24
9. Herbert PN, Assmann G, Gotto AM, Fredrickson DS (1983) Familial lipoprotein deficiency: abetalipoproteinemia, hypobetalipoproteinemia, and Tangier disease. In: Stanbury JB, Wyngaarden JB, Fredrickson DS, Goldstein JL, Brown MS (eds) The metabolic basis of inherited disease, 5th edn. McGraw-Hill, New York, pp 589–621
10. Leitersdorf E, Gottehrer N, Fainaru M, Friedlander Y, Friedman G, Tzivoni D, Stein G (1986) Analysis of risk factors in 532 survivors of first myocardial infarction hospitalized in Jerusalem. Atherosclerosis 59: 75–93
11. Muls E, Blaton V, Rosseneu M, Lesaffre E, Lamberigts G, De Moor P (1982) Serum lipids and apolipoproteins AI, AII and B in hyperthyroidism before and after treatment. J Clin Endocrinol Metab 55: 459–464
12. Muls E, Rosseneu M, Blaton V, Lesaffre E, Lamberigts G, De Moor P (1984) Serum lipids and apolipoproteins AI, AII and B in primary

hypothyroidism before and during treatment. Eur J Clin Invest 14: 12–15

13. Riesen WF, Mordasini R, Salzmann C, Theler A, Ourtner HP (1980) Apolipoproteins and lipids as discriminators of severity of coronary heart disease. Atherosclerosis 37: 157–162
14. Rosseneu M, Bury J (1988) Apolipoprotein assays for the diagnosis of hyperlipidemias. Impact on Prevention of Atherosclerotic Diseases: 143–154
15. Rosseneu M, De Backer G, Caster H, Hulstaert F, Burj J (1987) Distribution and composition of HDL subclasses in students whose parents suffered prematurely from a myocardial infarction in comparison with controls. Atherosclerosis 63: 231–235
16. Rosseneu M and Van Biervliet JP (1985) Screening and follow-up of infants with dyslipoproteinemia. In: Widhalm K, Naito H (eds) Detection and treatment of lipid and lipoprotein disorders in childhood. AR Liss, New York, pp 79–86
17. Schaefer EJ (1987) Clinical biochemical and genetic features in familial disorders of high density lipoprotein deficiency. Atherosclerosis 4: 303–315
18. Sedlis SP, Schechtman KB, Ludbrook PA, Sobel BE, Schonfeld G (1986) Plasma apoproteins and the severity of coronary artery disease. Circulation 73: 978–986
19. Sniderman AD, Shapiro S, Marpole D, Skinner B, Teng B, Kwiterovich PO (1980) Association of coronary atherosclerosis with hyperapobetalipoproteinemia (increased protein but normal cholesterol levels in human plasma low density lipoproteins). Proc Natl Acad Sci (USA) 77: 604–608
20. Van Biervliet JP, Vinaimont N, Caster H, Rosseneu M (1982) A screening procedure for dyslipoproteinemia in newborns. Apoprotein quantitation on dried-blood spots. Clin Chim Acta 120: 191–200

Author's address: Dr. M. Rosseneu, Department of Clinical Chemistry, A.Z. St-Jan, B-8000 Brugge, Belgium.

Apolipoproteins and coronary heart disease

J. P. Deslypere

Department of Endocrinology, University of Gent, Gent, Belgium

Summary

Clinical and experimental studies on apolipoprotein composition and atherogenicity are reviewed including case control, angiographic, prospective, and familial studies. Recent studies investigated the value of apolipoprotein measurements in the assessment of coronary hard disease risk. Most of the studies indicate that apo AI and B are strong CHD risk predictors. Apo B seems to be more strongly associated with cardiovascular risk than LDL-C or HDL-C levels. Another result was that the apolipoprotein levels are good indicators of the presene or absence of CHD. Apo B is a good indicator of the severity of disease.

Keywords: Review of studies, apolipoproteins, coronary heart disease, case control, angiographic, prospective and familial studies.

Introduction

There is ample evidence to relate blood lipid levels to the pathogenesis of atherosclerosis in human subjects. For years attention has primarily focused on the blood cholesterol and triglyceride (TG) levels, although an inverse relationship between the frequency of coronary heart disease (CHD) and high density lipoproteins (HDL) has been demonstrated as early as in 1951. The importance of these early observations has been fully acknowledged in the last two decades, where it was demonstrated that the HDL cholesterol (HDL-C) level was an important predictor of CHD independent of other known risk factors [25].

These results are of great importance to public health, but do not apply to a similar extent to the prediction of individual risk. The atherogenic risk in a given patient is indeed only partially characterized by his total cholesterol (TC) and HDL-C levels. Hence, other parameters are necessary to permit early detection of coronary heart disease risk. Therefore in more recent years, the protein moiety of the lipoproteins has been receiving increasing interest. Different investigations published at the end of the seventies suggested that the measurement of the apolipoproteins AI and B would be more valuable for the prediction of coronary risk on an individual basis than the determination of the classic lipid and lipoprotein parameters [2, 15, 30, 33]. This paper gives a review of the clinical and experimental studies on apolipoprotein composition and atherogenicity published since these first and very promising reports.

Studies on apolipoprotein composition and atherogenicity

Most studies have primarily analysed apo B and apo AI levels. Cut-off points for these apolipoproteins have generally been set at 130 mg/dl and 120 mg/dl respectively, by most authors [15].

In some studies apo AII has been measured, with normal values lying between 25 and 45 mg/dl.

More recently, attention has been focused on Apo E and Lp(a) levels. Normal serum values for Apo E vary between 1.5 and 9.0 mg/dl, while Lp(a) serum values are normally lower than 30 mg/dl.

The relation between apolipoprotein levels and CHD can be analysed by four sorts of studies: case control, angiographic, prospective and family studies [2, 17, 30, 33].

Case control studies

Many epidemiological and clinical studies carried out in the last decade have shown that apo AI and B are better markers for the risk of developing CHD than lipid and lipoproteins [2, 10, 11, 15, 30]. Most studies relating apolipoproteins to CHD have been of the case control design, with cases defined as subjects surviving a myocardial infarction (MI) or subjects having angina and angiographically proven CHD, with the control subjects matched for age, sex and sometimes other variables. Most of these studies report significantly lower apo AI levels and higher apo B levels in cases of CHD, compared to non-cases.

One of the first of these studies was published by our group. It showed that by using the HDL-C/TC and the apo AI/B ratios about 85% of either young MI survivors or healthy age- and sex-matched controls are correctly classified [11].

In a subsequent publication we have shown [10] that in a group of 70 male survivors of MI and in an equal number of healthy controls matched for sex, age and body mass index, the apo AI/B ratio discriminated best between cases and controls giving a 72% exact classification (Table 1).

This has subsequently been confirmed by others: Lehtonen et al. [23] found out that apo AI and apo AI/B ratio were the best parameters (± 70% success) to reclassify patients or controls to the correct groups, while in the study of Durrington et al. [12], apo B levels were the best single discriminating variable between MI survivors and controls (Fig. 1). Kostner [19] found out that MI survivors were classified more accurately by apo AI and B measurements than by LDL-C or HDL-C levels.

A possible confounding factor when considering the association of plasma HDL-C and apo AI levels with atherosclerotic disease is the triglyceride (TG) level. Indeed, in both cross-sectional and longitudinal studies, HDL-C has been found to be inversely related to TG levels [14]. TG levels are also more strongly inversely related to HDL-C than to apo AI. In some CHD case control studies in which the subjects were matched for plasma TG levels, the cases had still lower levels of both apo AI and HDL-C. In other studies

Table 1. Discriminative value of the lipid and apolipoprotein measurements based on the mean values [acc. to De Backer G, Rosseneu M, Deslypere JP (1982) Atherosclerosis 42: 197–203]

Variable	Mean value	Cases (> mean value)		Controls		Total % correctly classified	Chi-square
		n	%	n	%		
TC (mg/100 ml)	248	42	60	29	41	60	4.1*
HDL-C (mg/100 ml)	51	22	31	46	66	68	15.1***
Apo AI (mg/100 ml)	118	25	36	43	61	63	8.3**
Apo AII (mg/100 ml)	45	33	47	23	33	57	2.4[a]
Apo B (mg/100 ml)	128	49	70	18	26	72	25.4***
HDL-C/TC (%)	21	18	26	46	66	70	21.0***
Apo B/Apo AI (%)	114	43	61	12	17	72	26.9***
Apo AII/Apo AI (%)	38	41	59	17	24	68	15.6***

[a] p not significant; * $p<0.05$; ** $p<0.01$; *** $p<0.001$

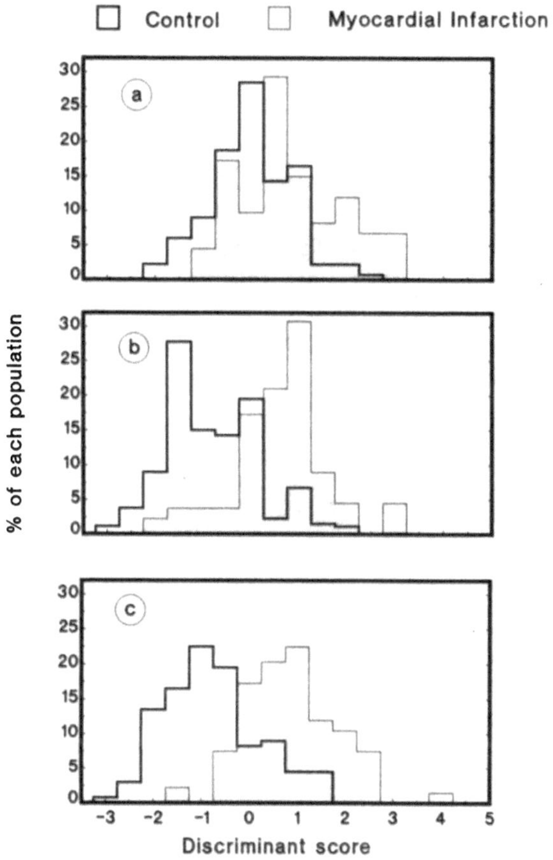

Fig. 1. Frequency distribution of the scores in controls and men in previous MI obtained by discriminant analysis using as variables (**a**) conventional risk factors (total serum cholesterol, serum triglycerides, serum HDL, cholesterol and parental history), (**b**) apolipoprotein B alone, and (**c**) variables selected by stepwise discriminant analysis (apo B, apo AI and parental history) [acc. to Durrington PN, Hant L, Ishola M, et al (1986) Br Heart J 56: 206–212]

where a significant difference in TG levels between cases and controls was found, discriminant analysis suggested an independent association of HDL-C or apo AI with CHD [25].

In almost all studies in which apo B levels have been examined, the cases had significantly higher levels of apo B than the controls. The majority of the studies [6, 8, 19, 25] also suggest that the apolipoprotein B level is a better predictor of CHD than TC or LDL-C.

The observation by Sniderman et al. [32] that the LDL-apo B concentration discriminates between subjects with or without CHD even when the LDL-C concentration is normal, has added to the interest in LDL-apo B measurements in the clinical setting.

Over-production of apo B which causes increased flux of LDL, overloading the reverse cholesterol transport from the arterial wall through affinity of apo B for the interstitial substance, has been suggested to cause the premature coronary atherosclerosis seen in normolipemic patients with elevated LDL-apo B levels.

High LDL-apo B concentrations have also been found in some, but not all, hypertriglyceridemic patients, a finding which might explain the controversy regarding hypertriglyceridemia as an independent risk factor for CHD [32].

Experiences from measurement of apo E are also sparse. Bittolo Bon et al. [5] measured apo E in phosphotungstate $MgCl_2$ supernatants of plasma from 40 normolipemic MI survivors and 40 control subjects. Although plasma total apo E did not differ between the two groups, the MI survivors had lower HDL-apo E concentrations. The reduction of HDL-apo E was greater and of a much higher statistical significance level than those of HDL-C or apo AI and AII. The authors conclude that these lower HDL apo E levels could be of significance for the reversed cholesterol transport.

Furthermore, it is known [1, 6] that the apolipoprotein E phenotype is an important diagnostic marker for type III-hyperlipoproteinemia and that the apolipoprotein E phenotype determines a significant proportion of the variance in plasma cholesterol and apo B levels. Indeed, the apolipoprotein phenotype E 4/4 and E 4/3 have proven to be associated with elevated TC, LDL-C and apo B levels. Since apo E gene frequencies vary considerably between different ethnic groups, the apo E gene locus thus contributes to differences in lipid levels between populations.

In addition to apo AI and B, there is growing evidence that Lp(a) is a good biochemical marker for both cardiovascular and cerebrovascular disease [20, 26, 29, 34]. This is especially true in patients with familial hypercholesterolemia. The increase in risk is independent of age, sex, smoking status and serum levels of TC, TG or HDL-C. The higher level is the result of genetic influence. It is now generally accepted that when levels of Lp(a) exceed 30 mg/dl (as occurs in about 20% of whites) the risk of CHD approximately doubles, while when both Lp(a) and LDL are elevated, the risk rises as much a five-fold.

Angiographic studies

Several recent studies have related the concentration of apolipoprotein to the extent and severity of CHD assessed by angiography [4, 16, 21, 24, 27, 28].

The study of Maciejko et al. [24] was the first to conclude that apo AI was by itself more useful than HDL-C for identifying patients with CHD (Fig. 2). This has subsequently been confirmed by Naito [27], Kottke et al. [21] and Reinhart et al. [28]. The data of Hamsten et al. [16] also demonstrated the importance of apo B concentrations (Table 2).

These studies were also confirmed by Barbir et al. [4] who showed that increased LDL-apo B and reduced apo AI concentrations were good discriminants between patients with angiographically confirmed CHD and controls (Fig. 3).

Prospective studies

There are only a few prospective studies on apolipoprotein concentrations and incidence of CHD.

The Tromso Heart Study [17] represents the first pro-

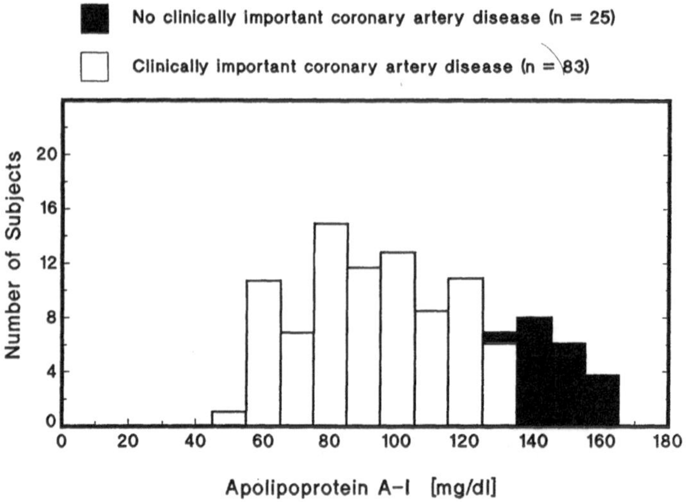

Fig. 2. Apo AI levels in patients with and without clinically important coronary artery disease [acc. to Maciejko JJ, Holmes DR, Kottke BA, et al (1983) N Engl J Med 309: 385–389]

Table 2. Correlation coefficients between lipids and coronary scores [acc. to Hamsten A, Walldius G, Szamosi A, et al (1986) Circulation 73: 1097–1110]

Parameters	Coronary atheromatosis score	Coronary stenosis score
TC	0.37*	0.08
TG	0.075	−0.024
LDL-C	0.40*	0.16
HDL-C	−0.14	−0.12
Apo AI	−0.078	−0.077
Apo B	0.43*	0.098
Apo B/AI	0.37*	0.17

* $p<0.01$

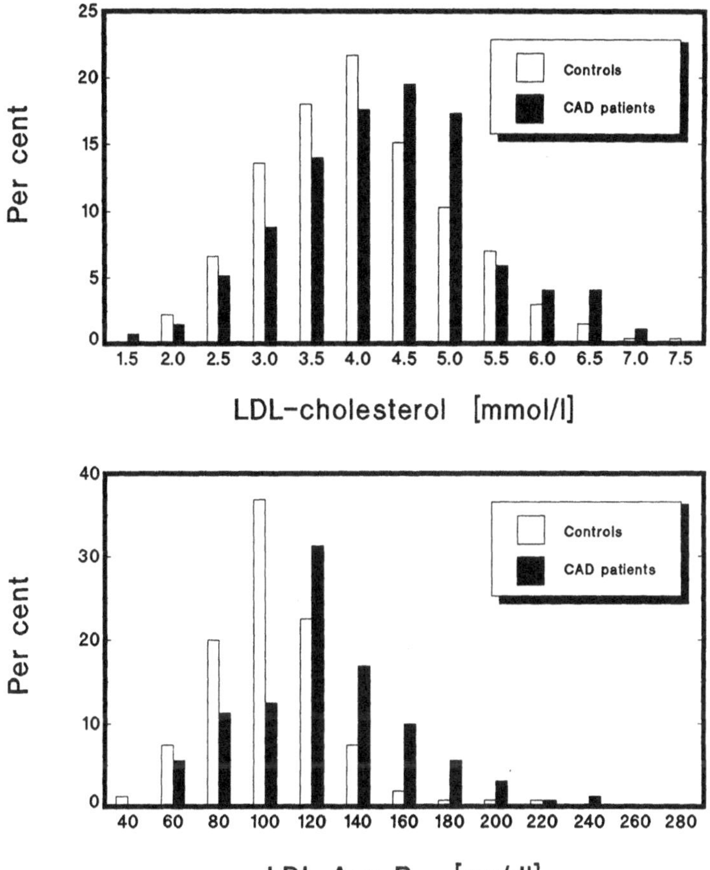

Fig. 3. Frequency distributions of LDL cholesterol in controls and patients with coronary artery disease (upper part) and frequency distributions of LDL-apo B in controls (n = 339) and patients with coronary artery disease (CAD patients) [acc. to Barbir M, Wile D, Trayner I, et al (1988) Br Heart J 60: 397–403]

spective study to determine the value of an apolipoprotein as a predictor of the risk of CHD. In a Finnish study by Salonen et al. [31] 92 persons (69 men and 23 women) with no previous MI who died from CHD during a 5-year follow-

Table 3. Relative risk of death from CHD. 5-year follow-up study [acc. to Salonen JP, Salonen R, Penttila I, et al (1985) Am J Cardiol 56: 220–231]

Parameters	All persons	Persons with no chest pain on effort
Apo AI <1.25 g/l	1.5	2.5
P/S ratio <0.28	3.5	5.6
Selenium <45 g/l	0.9	3.2

P/S ratio: polyunsaturated to saturated fatty acid ratio

Table 4. Lipids in off-spring of MI patients. Case control study [acc. to De Backer G, Hulstaert F, De Munck K, et al (1986) Am Heart J 112: 478–484]

Parameters	Men		Women	
	cases	controls	cases	controls
TC (mg/dl)	187	172	187	197
HDL-C (mg/dl)	51.6	52.0	54.6	62.1
Apo AI (mg/dl)	123*	138	134*	155
Apo B (mg/dl)	94.5	92.6	93.5	104.6
LDL-C (mg/dl)	119	105	118	120
Non-HDL-C / Apo B	1.43*	1.29	1.41*	1.26
HDL-C / Apo AI	0.42*	0.37	0.41	0.40

* $p<0.01$

up were compared to 92 control subjects. Low serum apo AI concentrations were associated with a 2.5-fold increased risk of death from CHD when serum cholesterol level and mean arterial blood pressure were controlled (Table 3).

Familial studies

Familial studies may help to elucidate the predictive value of the apolipoprotein parameters, without requiring many years of follow-up. Different studies have compared the levels of apolipoproteins in relatives of persons with CHD [9, 13, 22]. Almost all studies show a stronger association of parental history of CHD with apolipoprotein levels than with lipid or lipoprotein cholesterol. Apo B and apo AI levels are equally important.

Kukita èt al. [22] were one of the first to report that apo AI and apo B levels better distinguished relatives of CHD patients from controls than did HDL-C, TC and TG. After univariate analysis between CHD patients and controls, apo B was the best discriminator followed by HDL-C in men, while in women, apo AI was the best discriminator [23].

Freedman et al. [13] and De Backer et al. [9] subsequently confirmed that children whose fathers reported that they had suffered MI at a young age, had significantly lower levels of apo AI and lower ratios of LDL-C or non-HDL-C/apo B as well as significantly higher ratios of apo AI or HDL-C/apo AI than children whose fathers had not suffered a heart attack (Table 4). This was seen in both boys and girls [17]. In children whose mothers had reported having an MI, the apo B to apo AI ratios were higher [13]. Interestingly, serum lipoprotein cholesterol measurements were not related to the MI history of either the father or the mother [9, 13].

Conclusion

Most of the studies on apolipoprotein levels and CHD indicate that apo AI and apo B are strong CHD risk predictors, that apo B seems to be more strongly associated with cardiovascular risk than LDL-C or HDL-C levels and that the apolipoprotein levels are good indicators of the presence or absence of CHD. Apo B is also a good indicator of the severity of disease.

References

1. Assmann G (ed) (1989) Lipid metabolism disorders and coronary heart disease. MMV Medizin Verlag, München
2. Avogaro P, Bittolo Bon G, Cazzalato G (1979) Are apolipoproteins better discriminators than lipids for atherosclerosis? Lancet i: 901–903
3. Avogaro P, Bittolo Bon G, Cazzolato G, Rovai E (1980) Relationship between apolipoproteins and chemical components of lipoproteins in survivors of myocardial infarction. Atherosclerosis 37: 69–76
4. Barbir M, Wile D, Trayner I, Aber VR, Thompson GR (1988) High prevalence of hypertriglyceridemia and apolipoprotein abnormalities in coronary artery disease. Br Heart J 60: 397–403
5. Bittolo Bon G, Cazzolato G, Saccardi M, Kostner GM, Avogaro P (1984) Total plasma apo E and high density lipoprotein apo E in survivors of myocardial infarction. Atherosclerosis 53: 69–75
6. Boerwinkle E, Utermann G (1988) Simultaneous effects of the apolipoprotein E polymorphism on apolipoprotein E, apolipoprotein B and cholesterol metabolism. Am J Hum Genet 42: 104–112
7. Brunzell J, Mazzone T, Motulsky A, Albers JJ (1982) Abnormalities in high density lipoprotein composition in familial combined hyperlipidemia. Arteriosclerosis 2: 416–417
8. Brunzell JD, Albers JJ, Chait A (1983) Plasma lipoproteins in familial combined hyperlipidemic and monogenic familial hypertrigylceridemia. J Lipid Res 24: 147–155
9. De Backer G, Hulstaert F, De Munck K, Rosseneu M, Van Parijs L, Damaix M (1986) Serum lipids and apolipoprotein in students whose parents suffered prematurely from a myocardial infarction. Am Heart J 112: 478–484
10. De Backer G, Rosseneu M, Deslypere JP (1982) Discriminative value of lipids and apolipoproteins in coronary heart disease. Atherosclerosis 42: 197–203
11. Deslypere JP, De Backer G, Rosseneu M, Vermeulen A (1981) Lipid and apoprotein levels in myocardial infarction survivors: a case-control study. Acta Cardiol 27: 95–102
12. Durrington PN, Hant L, Ishola M, Kane J, Stephens WP (1986) Serum apolipoproteins AI and B and lipoproteins in middle aged men with and without previous myocardial infarction. Br Heart J 56: 206–212
13. Freedman DS, Srinivasan SR, Shear CL, Franklin FA, Webber LS, Berenson GS (1986) The relation of apolipoproteins AI and B in children to parenteral myocardial infarction. N Engl J Med 315: 721–726
14. Gordon DJ, Rifkind B (1989) High density lipoprotein – the clinical implications of recent studies. N Engl J Med 321: 1311–1316
15. Grundy S, Vega G (1990) Role of apolipoprotein levels in clinical practice. Arch Int Med 150: 1579–1582

16. Hamsten A, Walldius G, Szamosi A, Dahlen G, de Faire U (1986) Relationship of angiographically defined coronary artery disease to serum lipoproteins and apolipoproteins in young survivors of myocardial infarction. Circulation 73: 1097–1110
17. Ishikawa T, Fidge N, Thelle DS, Forde OH, Miller NE (1978) The Tromso Heart Study. Serum apolipoprotein AI concentration in relation to future coronary heart disease. Eur J Clin Invest 8: 179–182
18. James RW, Martin B, Pometta D, Grab D (1986) Apoprotein D in a healthy male population and in male myocardial infarction patients and their male first degree relatives. Atherosclerosis 60: 49–53
19. Kostner GM (1983) Apoproteins and lipoproteins of human plasma, significance in health and in disease. Adv Lipid Res 20: 1–43
20. Kostner GM, Avogaro P, Cazzolato G (1981) Lipoprotein Lp(a) and the risk for myocardial infarction. Atherosclerosis 38: 51–61
21. Kottke BA, Zinsmeister A, Holmes D, Kneller R, Hallaway B, Mao S (1986) Apilipoproteins and coronary artery disease. Mayo Clin Proc 61: 313–320
22. Kukita H,l Hiwada K, Kokubu T (1984) Serum apolipoprotein AI, AII and B levels and their discriminative values in relatives of patients with coronary artery disease. Atherosclerosis 51: 261–267
23. Lehtonen A, Marniemi J, Inberg M, Maatela J, Alanen E, Nuttymaki K (1986) Levels of serum lipids, apolipoproteins apo AI and B and pseudocholinesterase activity and their discriminative value in patients with coronary bypass operation. Atherosclerosis 59: 215–221
24. Maciejko JJ, Holmes DR, Kottke BA, Zinsmeister AR, Dinh DM, Mao SJ (1983) Apolipoprotein AI as a marker of angiographically assessed coronary artery disease. N Engl J Med 309: 385–389
25. Miller NE (1987) Associations of high lipoprotein subclasses and apolipoproteins with ischemic heart disease and coronary atherosclerosis. Am Heart J 113: 589–597
26. Murai A, Miyahara T, Fujimoto N (1986) Lp(a) lipoprotein as a risk factor for coronary heart disease and cerebral infarction. Atherosclerosis 59: 199–204
27. Naito HK (1985) The association of serum lipids, lipoproteins and apolipoproteins with coronary artery disease assessed by coronary arteriography. Ann NY Acad Sci 454: 230–238
28. Reinhart R, Gani K, Arndt M, Broste S (1990) Apolipoproteins AI and B as predictors of angiographically defined coronary artery disease. Arch Int Med 150: 1629–1633
29. Rhoads GG, Dahlen G, Berg K (1986) Lp(a) lipoprotein as a risk factor for myocardial infarction. JAMA 256: 2540–2544
30. Riesen WF, Mordasini R, Salzmann C, Theler A, Gurtner MP (1980)

Apoproteins and lipids as discriminators of severity of coronary heart disease. Atherosclerosis 37: 157–162
31. Salonen JP, Salonen R, Penttila I (1985) Serum fatty acids, apolipoprotein, selenium and vitamin antioxidants and the risk of death from coronary artery disease. Am J Cardiol 56: 226–231
32. Sniderman AD, Wolfson C, Teng B, Franklin FA, Bachorik PS, Kwiterovich PO (1982) Association of hyperapobetalipoproteinemia with endogenous hypertriglyceridemia and atherosclerosis. Ann Intern Med 97: 833–839
33. Wiklund O, Fager G, Olofsson SO, Wilhelmsson C, Bondgers G (1980) Serum apolipoprotein levels in relation to acute myocardial infarction: determination of apolipoprotein B. Atherosclerosis 37: 631–636
34. Zenker G, Koltringer P, Bone G (1986) Lipoprotein(a) as a strong indicator for cerebrovascular disease. Stroke 17: 942–945

Author's address: Dr. J. P. Deslypere, Department of Endocrinology, University of Gent, B-9000 Gent, Belgium.

Apolipoproteins in children with familial history of myocardial infarction

K. Widhalm

Department of Pediatrics, University of Vienna, Vienna, Austria

Summary

There is a great interest whether offsprings from families with manifest coronary heart disease can be discriminated from children without any family history. For this reason we studied serum lipids, lipoproteins and apoproteins in 291 offsprings from 152 fathers who had a myocardial infarction before the age of 55 years, in comparison with 283 healthy age matched controls. Statistical analysis reveals marked differences between the groups, showing that HDL-cholesterol, LDL-cholesterol and total cholesterol/HDL-cholesterol are the best discriminators. However, apolipoproteins AI and B are very good discriminators in the group < 20 years and the group > 20 years, respectively. The ratio AI/B is different in both age groups ($p < 0.01$ vs. 0.05).

However, there is no evidence that apolipoproteins add more information in order to discriminate between risk and non-risk families. More long-term follow-up seems to be necessary and appropriate statistical models should be used in order to support the hypothesis that apoproteins are better indicators for later cardiovascular risk than lipoproteins.

Keywords: Apolipoproteins in children, myocardial infarction, cardiovascular disease, risk factors, risk and non-risk families.

Introduction

It is generally accepted that the atherosclerotic process begins in early life with the accumulation of lipid, principally cholesterol and its esters, in the intima of the large elastic and muscular arteries to form fatty streaks. At some sites in the coronary arteries and other vessels, fatty streaks thicken by continued accumulation of abnormal lipid, smooth muscle cells and connective tissue to form fibrous plaques. In young adulthood, fibrous plaques begin to increase in extent and thickness and undergo calcification and vascularization.

Intimal lipid deposits and young children are primarily intracellular. Extracellular lipid, although present, does not accumulate and does not impair normal intimal structure. Such lesions, sometimes visible as fatty streaks, make their initial appearance around age 10 in the coronary arteries. The majority of those who are to have coronary atherosclerosis develop their initial lesions at puberty.

Risk factors for cardiovascular disease

The association between the occurrence of early atherosclerotic lesions and the antemortem risk factors for cardiovascular disease has been shown clearly by the study performed on adolescents who deceased by accidents in New Orleans. Thus, a significant relationship between the atherosclerotic fatty streak involvement of the aorta and the antemortem estimated LDL-C levels has been shown [7, 12] (Fig. 1).

Furthermore, a significant association between the percentage of total surface involvement of the coronary arteries with fatty streaks and the VLDL concentration could be shown. Thus, although within a relatively small group of subjects, there exists some evidence that the concentration of certain lipoproteins is more directly associated to clinical disease than the corresponding concentrations of lipoprotein cholesterol fractions alone [2]. In addition, elevated levels of LDL apo B without an increase of LDL-C – a condition termed hyperapobetalipoproteinemia – may be associated with the development of cardiovascular disease [10]. Furthermore, a decreased ratio of LDL-C/LDL-apo B has been found in patients with coronary artery disease [10] and in

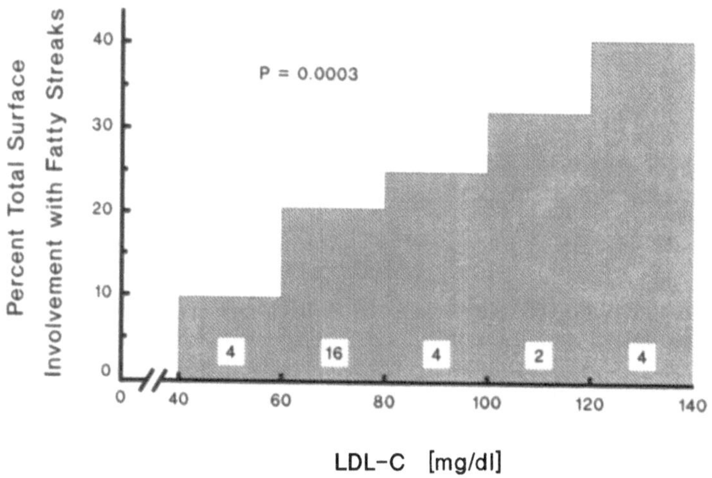

Fig. 1. Atherosclerotic fatty-streak involvement of the aorta related to levels of low-density lipoprotein cholesterol (LDL-C) in 30 young persons

the children of persons with premature myocardial infarction (MI) and hyperapobetalipoproteinemia.

Offspring from "risk" and "non-risk" families

There are few studies with children of families with premature MI investigating whether it is possible to discriminate between a risk group from a non-risk group measuring lipoproteins and apolipoproteins [3, 4].

We are reporting on a study which has been performed on 152 fathers who had an MI under the age of 55 years. The diagnosis of MI has been made by means of typical clinical symptoms, estimation of enzyme pattern and has been confirmed in the majority of the cases by coronary angiography.

Blood was drawn from the patients several weeks after the MI during a steady state metabolic situation. In total, blood was drawn from 291 offsprings of the patient at a fasting state; most of them came to our clinic, some of them sent the sample by mail to our laboratory.

In addition, 283 healthy children matched for age from a school cholesterol screening program served as controls. Blood was drawn during the same half year together with the children from the risk group.

Methods

Cholesterol and TG were measured by means of enzymatic methods (Boehringer Mannheim, Germany).

HDL-C was determined by a polyanion precipitation method (Immuno Diagnostics, Vienna).

LDL-C and VLDL-C were measured by the Lipid Research Clinics technique, using Beckman L 7 Ultracentrifuge and polyethyleneglycol for precipitation of LDL. Apo AI and B were measured by a photometric method (Boehringer Mannheim, Germany). For quality control, apolipoprotein standard (from lyophilized human sera) and Precinorm® L were used.

Apo AII was determined by radioimmunodiffusion plates from Immuno AG Vienna, using the reference standard (apolipoprotein human) and the Norm Control Apolipoprotein Human Serum. All estimated coefficients of variations were within the 5% range.

Table 1. Mean values (mg/dl ± SD) of fathers with premature myocardial infarction (before the age of 55)

Parameter	Mean	± SD	n
Age	46.7	6,9	152
TC	236.4	52.9	141
TG	170.2	96.9	141
HDL-C	38.8	11,4	141
LDL-C	180.2	49.5	141
VDL-C	14,2	10,1	141
Apo A1	129.5	25,8	129
Apo A2	39.0	12,6	124
Apo B	117.1	30.0	130
TC/HDL-C	6,5	2,11	141
TC/LDL-C	1,33	0.14	141
LDL-C/HDL-C	5.00	1,92	141
Apo A1/Apo B	1,18	0.36	128
LDL-C/Apo B	0.67	0.13	130
HDL-C/Apo A1	0.30	0.06	129

Results

The results of TC, TG, LDL-C, HDL-C and apo AI, AII and the ratios between TC and HDL-C, LDL-C and HDL-C and the apolipoproteins are displayed in Table 1. As can be seen, the mean age of the subjects who had suffered from an MI was relatively young (46.7 years), the mean levels for

Table 2. Mean values (± SD) and significances between healthy controls and offsprings from families with premature myocardial infarction

Age Group < 20 Yrs			
Parameter	Patients (n=174)	sign.	Controls (n=168)
Age	12.7±4.8	n.s.	12.5±3.4
TC	180.2±49.5	n.s.	166.8±25.9
TG	84.1±55.4	p<0.05	90.5±47.4
HDL-C	47.2±11.3	p<0.001	51.7±10.6
LDL-C	126.9±47.7	p<0.005	108.1±22.9
VDL-C	6.6±5.7	p<0.005	7.4±4.3
Apo A1	140.1±23.8	p<0.001	153.5±23.4
Apo A2	42.5±11.7	n.s.	44.8±7.2
Apo B	82.7±24.9	p<0.001	67.8±16.4
TC/HDL-C	3.98±1.26	p<0.001	3.33±0.73
LDL-C/Apo B	2.85±1.26	p<0.001	2.17±0.63
Apo A1/Apo B	1.84±0.60	p<0.001	2.39±0.64
LDL-C/Apo B	0.72±0.23	p<0.005	0.64±0.15
HDL-C/Apo A1	0.34±0.07	n.s.	0.34±0.06

Age Group > 20 Yrs			
Parameter	Patients (n=117)	sign.	Controls (n=115)
Age	26.2±5.0	n.s.	26.1±4.6
TC	204.7±57.1	p<0.05	185.2±35.7
TG	136.0±114.0	n.s.	104.2±50.9
HDL-C	47.6±12.5	p<0.05	51.0±11.5
LDL-C	147.4±51.9	p<0.005	127.0±33.2
VDL-C	10.6±10.7	n.s.	7.9±4.2
Apo A1	150.1±27.8	n.s.	150.6±22.0
Apo A2	44.3±11.5	p<0.05	46.7±7.6
Apo B	94.3±29.3	p<0.05	83.6±21.5
TC/HDL-C	4.54±1.66	p<0.001	3.76±0.99
LDL-C/Apo B	3.29±1.36	p<0.001	2.45±0.74
Apo A1/Apo B	1.73±0.63	p<0.05	1.93±0.64
LDL-C/Apo B	0.67±0.16	p<0.05	0.61±0.10
HDL-C/Apo A1	0.32±0.05	p<0.05	0.34±0.04

TC and LDL-C high, whereas the HDL-C levels were considerably low.

The lipid, lipoprotein and apolipoprotein data for the offspring of the patients compared to age matched controls, are given in Table 2. Due to age-related reference levels, the children were divided into 2 groups, younger and older than 20 years. As can be seen, there is a clear tendency for higher levels in the offspring of fathers with MI for TC, TG, LDL-C, VLDL-C and apo B. Statistical significant different mean values could be found for LDL-C, apo B and the ratios TC/HDL-C, LDL-C/HDL-C and the ratios AI/B, LDL-C/apo B in both groups. Furthermore, the HDL-C levels are significantly lower in the offspring than the controls.

Discussion

Our study shows that children from fathers who had an MI at young age (< 50 years) differ significantly with regard to the lipoprotein and apolipoprotein levels. Among these parameters LDL-C and the ratio TC/HDL-C seem to be the best discriminators between offspring from "risk" families and "non-risk" families. Apo B and the ratios LDL-C/HDL-C, apo AI/B and LDL-C/apo B distinguished well between these two groups. These findings extend the results of several clinical studies in adults and some studies in children and adolescents.

Sveger and coworkers reported on a study based on questions that out of 84 families with a history of cardiovascular disease, 12 children (14%) had elevated LDL-C levels [13]. Lee and coworkers published that 51% of the progeny of fathers with MI had elevated TG, LDL-C or diminished HDL-C [5]. On the other hand, de Baecker et al. and recently Perova et al. reported that apolipoproteins (AI, AII) or the ratio apo B/AI seem to be better discriminators than lipoprotein-lipid levels alone [3].

From our data it can be concluded that in the offspring of fathers with premature MI, irrespective of their age, LDL-

C and the TC/HDL-C ratio are the strongest discriminators which can differentiate them from healthy children without any family history. Furthermore, it can be clearly demonstrated that apo AI and B are also different, which could be of great importance for epidemiological studies. Further long-term follow-up studies and statistical models seem to be necessary in order to support the previously published hypothesis which indicates that apolipoproteins in particular the apo B/AI ratio adds more information with regard to a later cardiovascular risk.

References

1. Avogaro P, Bittolo GB, Cazzolato G, Grimel GB, Belluzzi F (1978) Plasma levels of apolipoprotein AI and apolipoprotein B in human atherosclerosis. Artery 4: 385–394
2. Brunzell JB, Sniderman AD, Albers JJ, Kwiterovich PO Jr (1984) Apoproteins B and AI in coronary artery disease in humans. Arteriosclerosis 4: 79–83
3. De Baecker G, Hulstaert F, De Munck K, Rosseneu M, Van Parijs L, Dramaix M (1986) Serum lipids and apoproteins in students whose parents suffered prematurely from a myocardial infarction. Am Heart J 112: 478–484
4. Freedman DS, Srinivasan SR, Shear CL, Franklin FA, Webber LS, Berenson GS (1986) The relation of apolipoproteins AI and B in children to parental myocardial infarction. N Engl J Med 315: 721–726
5. Lee J, Lauer RM, Clarke WR (1986) Lipoproteins in the progeny of young men with coronary artery disease: children with increased risk. Pediatrics 78: 330–337
6. Morrison JA, Khoury P, Melles MS, Kelly KA, Glueck CJ (1980) Identifying CHD risk factors in children: intrafamilial lipoprotein correlations. Prev Med 9: 484–495
7. Newman WP, Freedmann DS, Cresanta JL, Williamson GD, Webber LS, Berenson GS (1986) Relation of serum lipoprotein levels and systolic blood pressure to early atherosclerosis. N Engl J Med 314: 138–144
8. Schmidt GB, Wasserman AG, Muesing RA, Schlesselman SE, Larosa JC, Ross AM (1984) Lipoprotein and apolipoprotein levels in angiographically defined coronary artery disease in humans. Arteriosclerosis 4: 79–83

9. Schrott HG, Clarke WR, Wiebe DA (1979) Increased coronary mortality in relatives of hypercholesterolemia school children. The Muscatine Study. Circulation 59: 320–326
10. Sniderman A, Shapiro S, Marpole D, Skinner B, Tang B, Kwiterovich PO Jr (1980) Association of coronary atherosclerosis with hyperapobetalipoproteinemia (increased protein but normal cholesterol levels in human plasma low density (β) lipoproteins). Proc Natl Acad Sci USA 77: 604–608
11. Sniderman AD, Teng B, Genest J, et al (1985) Familial aggregation and early expression of hyperapobetalipoproteinemia. Am J Cardiol 55: 291–295
12. Stary HC (1989) Evolution and progression of atherosclerotic lesions in coronary arteries of children and young adults. Arteriosclerosis [Suppl 1, 9]: 1–19; 1–32
13. Sveger T, Fex G, Borgas N (1987) Hyperlipidemia in school children with family histories of premature coronary heart disease. Acta Paediatr Scand 76: 311–315

Author's address: Dr. K. Widhalm, Department of Pediatrics, University of Vienna, A-1097 Vienna, Austria.

Apolipoproteins AI and B in health and metabolic disease

H. Drexel, Th. Hopferwieser, and **J. R. Patsch**

Division of Clinical Atherosclerosis Research, Department of Medicine, University of Innsbruck, Innsbruck, Austria

Summary

The new turbidimetric methods were evaluated in a total of 175 patients including 59 subjects without metabolic disease, 27 with type I diabetes, 20 with type II diabetes, 25 with hypertension, 23 with hypercholesterolemia and 21 with hypertriglyceridemia. When compared to the nephelometric reference method, both apo AI and apo B showed excellent correlation coefficients (0.99). A good correlation was also found for both apo AI vs HDL-cholesterol ($r = 0.87$) and for apo B vs LDL-cholesterol ($r = 0.89$).

In type II diabetes and hypertriglyceridemia, apo AI was decreased to a lesser extent than HDL-cholesterol which is interpreted as a predominant decrease of HDL_2. Apo B was found elevated in the cohorts with type II diabetes, hypercholesterolemia, and hypertriglyceridemia reflecting an increased concentration of apo B-containing lipoproteins in these patients.

Keywords: Apolipoproteins, lipoprotein status, comparison with LDL-C and HDL-C, metabolic disease, diabetes, hypertension, hypercholesterolemia.

Introduction

The new turbidimetric methods for the determination of apo AI and B in plasma offer a rapid information on plasma lipoproteins in the clinical laboratory. At this stage of the introduction of a new assay, we wanted to investigate three questions:

– How does the new assay correlate to an established method of plasma apolipoprotein measurement (nephelometry)?

– When compared to LDL-cholesterol (LDL-C) and HDL-cholesterol (HDL-C), what are the advantages and disadvantages of the new methods; is there additional information on the patients' lipoprotein status?

– Are there specific considerations for normal probands and for patients with metabolic disease, such as diabetes (type I and type II), hypercholesterolemia, hypertriglyceridemia and hypertension?

Materials and methods

All assays were performed in duplicate on a COBAS® Mira Autoanalyzer (Roche, Basle, Switzerland).

For purposes of comparison, we measured apo AI and apo B by nephelometry also on a Beckman Array System (Beckman Instruments, Berkeley, CA, USA). Cholesterol and triglycerides (TG) were determined as described previously [1]. HDL-C was measured after precipitation of VLDL and LDL by dextrane sulfate [4]. C-peptide was measured by radioimmunoassay (RIA-mat, C-peptide II, Byk-Sangtec Diagnostica, Dietzenbach, Germany).

Patients

We investigated a total of 175 patients. Of these, 59 had no metabolic disease; 47 had diabetes mellitus according to the criteria of the National Diabetes Data Group [3], 27 had type I diabetes, defined by a negative reading for C-peptide, and 20 had type II diabetes. The age of the type I patients ranged from 20 to 64 years, mean was 31. Thirteen were on intensified insulin treatment with multiple daily injections, and 14 on conventional insulin therapy. The age of the type II patients ranged from 33 to 74 years, mean was 56. Twelve were on insulin, 5 on sulfonylurea, and 3 on diet-only regimens. The study also included 25 patients with hypertension, 23 patients with a plasma cholesterol exceeding 240 mg/dl (6.2 mmol/l) and 21 with TG in excess of 200 mg/dl (2.3 mmol/l).

Results and discussion

Method comparison for apo AI and apo B

In Fig. 1, the upper panel shows the data for apo AI, on the abscissa are the reference values, the readings on the ordinate refer to the new method using a COBAS® Mira Autoanalyzer. For the total of 175 patients, the correlation was excellent with a coefficient of 0.99. Slope was slightly < 1 with a positive intercept on the ordinate of 0.18 g/l. This means that lower apo AI values were slightly overestimated while higher values were slightly underestimated by the new method when compared to the nephelometric method.

The lower panel describes apo B. For our group of 175 patients the correlation coefficient was 0.99, slope 0.93 and the intercept was virtually zero.

Thus, in conclusion, the new method is entirely satisfactory for both apo AI and for apo B when compared to our laboratory routine method.

Comparison of apo AI and B with HDL-C and LDL-C

Here, we compared the readings of apo AI to those of HDL-C. Also, LDL-C was calculated according to Friedewald's formula [2] and compared to readings of apo B.

Table 1 and Fig. 2 depict the correlation between HDL-C and apo AI. For the total of 175 patients the correlation coefficient was 0.87; reasonably good if one considers that two HDL components were compared, whose ratio differs within the HDL particle spectrum ranging from HDL-2 to HDL-3.

The correlation between LDL-C and apo B (Table 1) was equally good with a correlation coefficient of 0.89. We herewith conclude that apo AI gives a good correlation with HDL-C as does apo B with LDL-C.

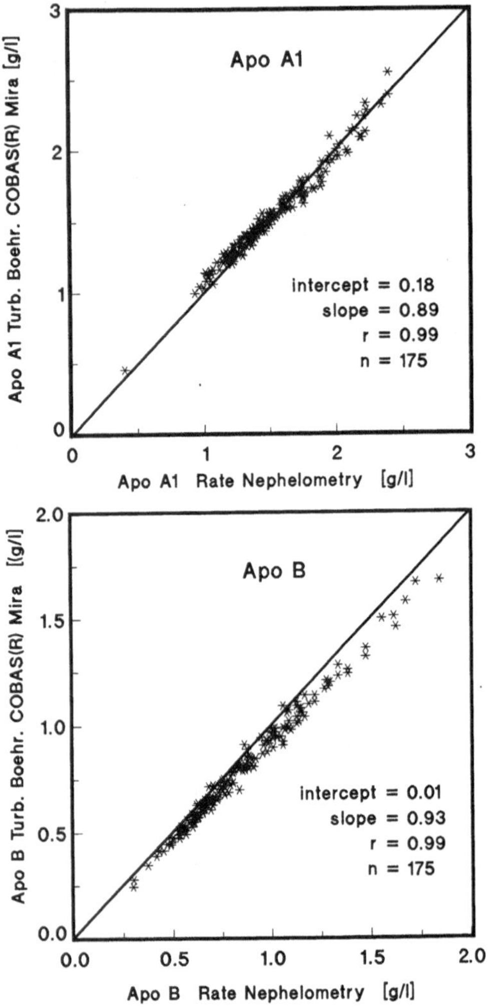

Fig. 1. Method comparison: *Top*: apo AI, turbidimetry/nephelometry; *bottom*: apo B, turbidimetry/nephelometry

Apo AI and apo B in health and metabolic disease

Table 2 shows the correlations between HDL-C and apo AI within the above defined subgroups. As was shown in Table 1

Table 1. Apo AI, apo B, HDL-C, LDL-C, method comparison for healthy and diseased individuals on different instruments

Method comparison

Instruments:
COBAS® Mira (turb.)
Beckman Array (neph.)

Methods:
Apo AI/B, Boehringer Mannheim, cholesterol
Apo AI/B, routine assays

Methods	n	Range		Intercept	Slope	r	Median of rel. differences
		min	max				
x = apo AI routine [g/l] y = apo AI BM	175	0.40	2.55	0.18	0.89	0.99	1.70
x = HDL-cholesterol [mmol/l] y = apo AI BM	175	0.28	3.38	0.70	0.87	0.87	−3.10
x = apo B routine [g/l] y = apo B BM	175	0.25	1.85	0.01	0.99	0.99	−6.20
x = LDL-cholesterol [mmol/l] y = apo B BM	158	0.25	7.51	0.07	0.89	0.89	−74.4

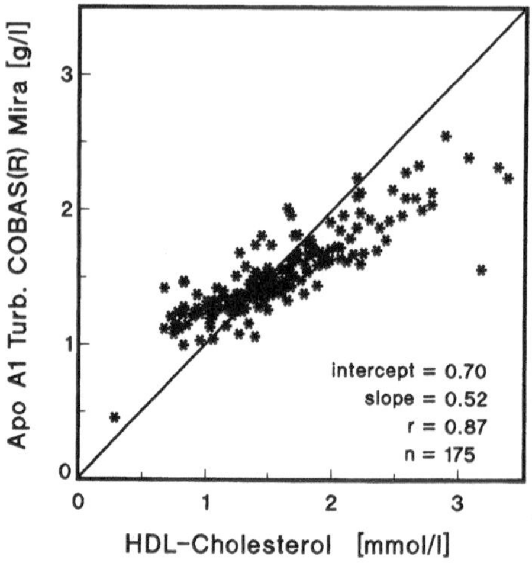

Fig. 2. Correlation of apo AI to HDL-C

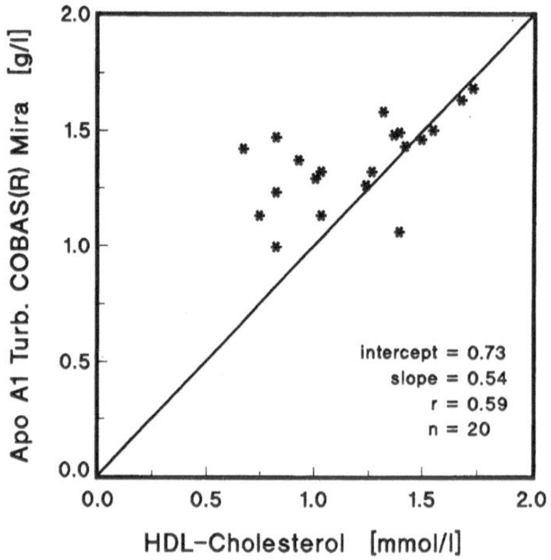

Fig. 3. Correlation between apo AI and HDL-C in type II diabetes

for all 175 patients the correlation coefficient was 0.87 and the slope was about 0.5. This also held true for the normal cohort, type I diabetics, hypertonics and hypercholestero-

Table 2. HDL-C and apo AI, method comparison for healthy and diseased individuals

Method comparison							
Instrument: COBAS® Mira (turb.)				Methods: x = HDL-cholesterol y = apo AI Boehringer Mannheim			
Patient groups	n	Range [mmol/l]		Inter- cept	Slope	r	Median of rel. differences
		min	max				
Normal persons	59	1.08	2.79	0.62	0.53	0.90	−12.0
Diabetics type I	27	0.90	2.39	0.62	0.53	0.94	−10.0
Diabetics type II	20	0.67	1.73	0.73	0.54	0.59	8.9
Hypertonic patients	25	0.83	2.68	0.60	0.60	0.94	3.6
Triglycerides >2.3 mmol/l	23	0.28	2.24	0.50	0.82	0.91	31.2
Cholesterol >6.2 mmol/l	21	0.96	3.30	0.42	0.70	0.77	−3.90

Table 3. LDL-C and apo B, method comparison for healthy and diseased individuals

Method comparison							
Instrument: COBAS® Mira (turb.)				Methods: x = LDL-cholesterol y = apo B Boehringer Mannheim			
Patient groups	n	Range [mmol/l]		Inter-cept	Slope	r	Median of rel. differences
		min	max				
Normal persons	59	0.25	3.74	0.11	0.21	0.75	−74.7
Diabetics type I	27	0.83	4.75	0.12	0.20	0.93	−74.2
Diabetics type II	15	0.63	6.35	0.19	0.21	0.84	−72.4
Hypertonic patients	23	0.47	5.13	0.14	0.21	0.93	−75.0
Trigylcerides >2.3 mmol/l	13	0.66	6.84	0.29	0.21	0.84	−68.8
Cholesterol >6.2 mmol/l	21	0.72	7.51	0.29	0.18	0.69	−75.5

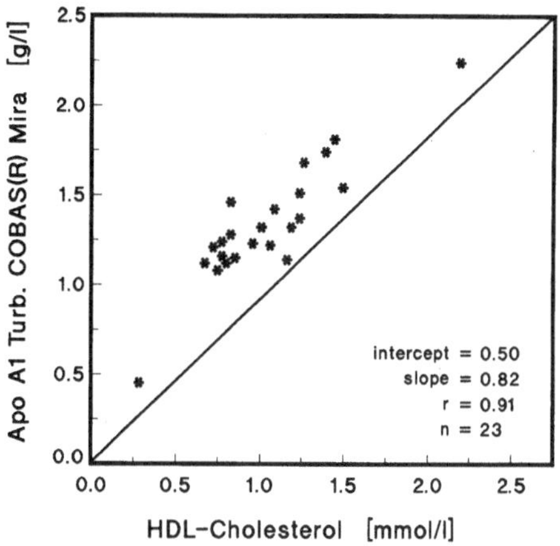

Fig. 4. Correlation between apo AI and HDL-C in subjects with increased plasma triglycerides (> 2.3 mmol/l)

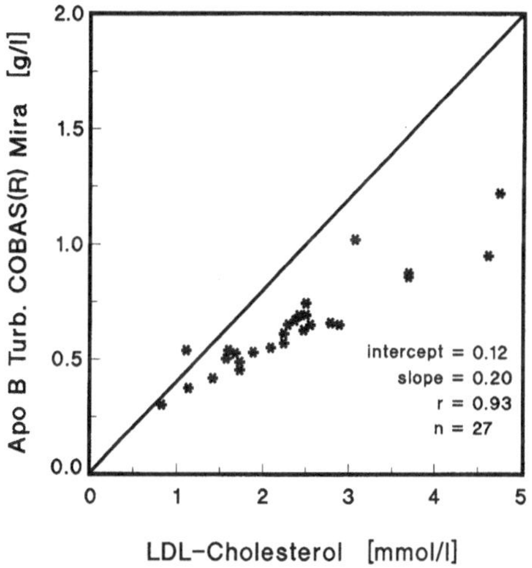

Fig. 5. Correlation between apo B and LDL-C in healthy individuals

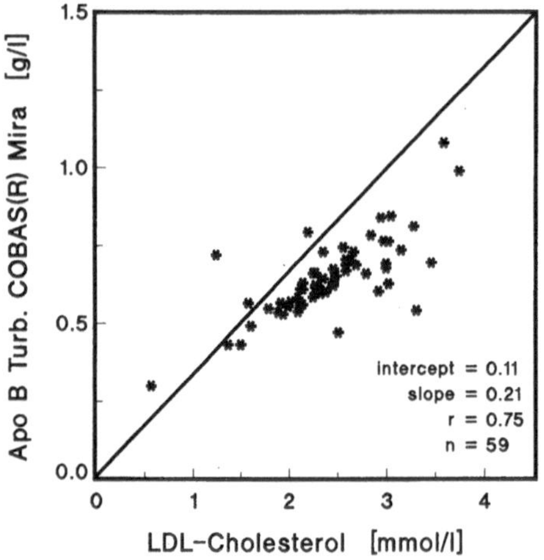

Fig. 6. Correlation between apo B and LDL-C in type I diabetes

lemic patients. However, the correlation coefficient was only 0.59 in type II diabetics. Figure 3 shows the plot for this subgroup. It is evident that some patients had lower HDL-C, but not lower apo AI. This could point to a selective loss of the more cholesterol-rich HDL-2 in some type II diabetic patients, resulting in a relative excess of the more protein-rich HDL-3. This is in excellent agreement with the fact that up to 50% of type II diabetics are hypertriglyceridemic with extremely low HDL-2 levels.

Hypertriglyceridemic patients (Fig. 4) also displayed a similar characteristic. All individuals were located left of the slope-1.0-line indicating that their HDL was relatively enriched in protein and relatively deficient in cholesterol. This finding agrees well with the demonstrated fact that in hypertriglyceridemia, HDL-2 particles are very low and the more protein-rich HDL-3 are enriched in triglycerides at the expense of cholesterol because of the action of the cholesterol-ester transfer protein pathway [5].

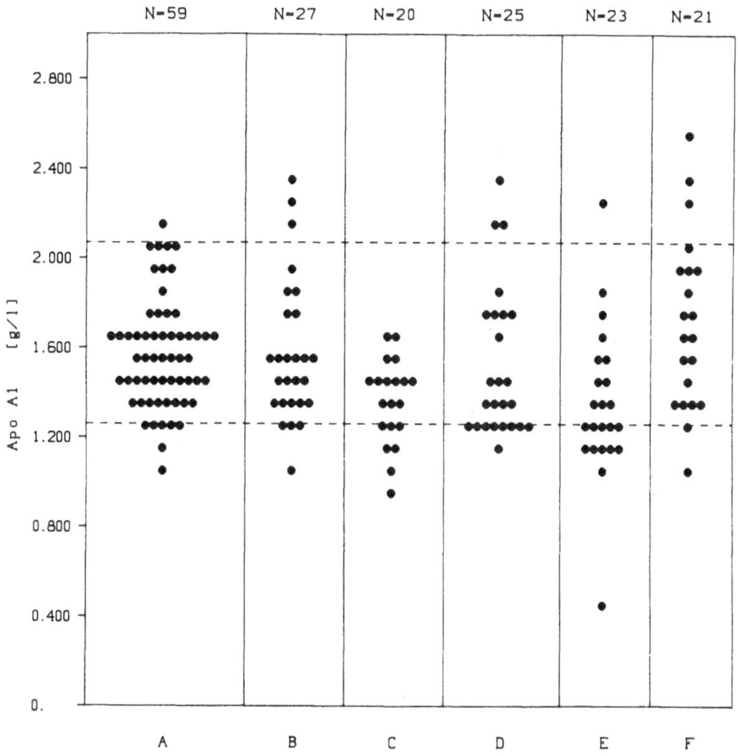

Fig. 7. Apo AI concentration in 6 patient groups. Selection of groups: *A* healthy individuals, *B* type I diabetics, *C* type II diabetics, *D* hypertonic patients, *E* plasma triglycerides > 2.3 mmol/l, *F* plasma cholesterol > 6.2 mmol/l

Table 3 shows the data for apo B versus LDL-C in these subgroups. The slope was practically identical in all groups, but the correlation coefficient, for example, was 0.75 in normal individuals versus 0.93 in type I diabetics. These two cohorts are shown in Fig. 5 and 6. The only difference appeared to be a smaller scatter in type I diabetics, which could mean that chemical composition was less variable in this cohort than in normal individuals.

After these analyses within subgroups, we looked at dif-

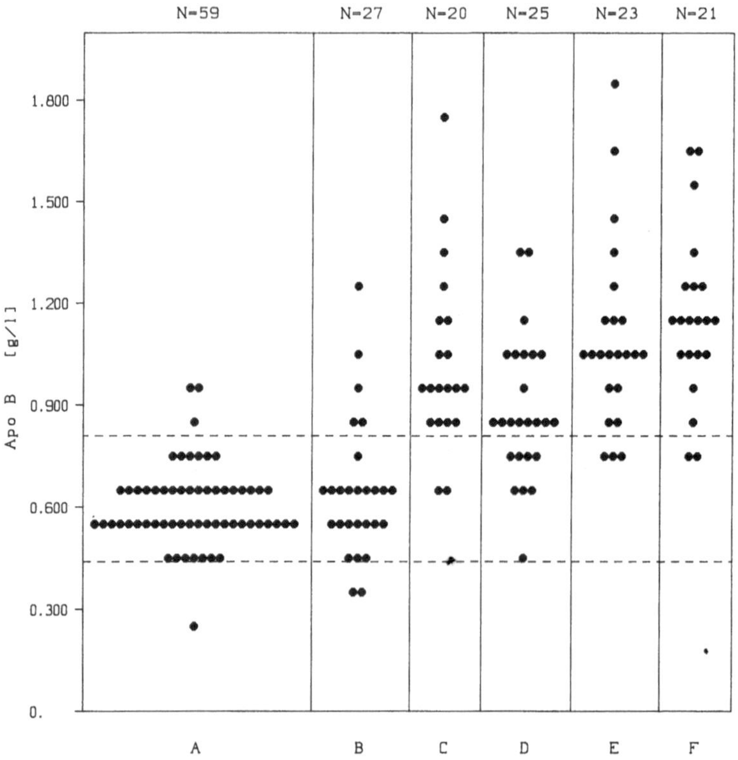

Fig. 8. Apo B concentration in 6 patient groups. Selection of groups as in Fig. 7

ferences between the subgroups. Triglycerides (data not shown) were elevated in type II diabetes, in hypertension and, by definition, in group E (hypertriglyceridemia). Figure 7 shows the distribution of apo AI in the different groups. Apo AI was low in type II diabetes and in hypertriglyceridemia and agreed well with the low HDL-C levels in type II diabetes and hypertriglyceridemia (data not shown). Apo B (Fig. 8) was elevated in type II diabetes, hypertriglyceridemia and, of course, in hypercholesterolemia reflecting the increased concentration of apo B-containing lipoproteins in these disorders.

Acknowledgements

This work was supported by grant HL-27341 from the National Institute of Health and by grant S46/06 from Austrian Fonds zur Förderung der Wissenschaftlichen Forschung. The authors wish to thank G. Tröbinger, J. Rössler, I. Pritz, I. Hauser and C. Comploy for excellent technical assistance.

References

1. Drexel H, Hopferwieser Th, Braunsteiner H, Patsch JR (1988) Effects of biosynthetic human proinsulin on plasma lipids in type II diabetes mellitus. Klin Wschr 66: 1171–1174
2. Friedewald WT, Levy RI, Fredrickson DS (1972) Estimation of the concentration of LDL-cholesterol in plasma without use of the preparative ultracentrifuge. Clin Chem 18: 499–506
3. National Diabetes Data Group (1979) Classification and diagnosis of diabetes mellitus and other categories of glucose intolerance. Diabetes 28: 1039–1057
4. Patsch W, Brown SA, Gotto AM, Patsch JR (1989) Evaluation of a dual precipitation method for measurement of cholesterol in high density lipoprotein subfractions HDL-2 and HDL-3 in human plasma. Clin Chem 36: 265–270
5. Patsch JR, Prasad S, Gotto AM, Patsch W (1987) High density lipoprotein$_2$. Relationship of the plasma levels of this lipoprotein species to its composition, to the magnitude of postprandial lipemia, and to the activities of lipoprotein lipase and hepatic lipase. J Clin Invest 80: 341–347

Authors' address: Dr. H. Drexel, Division of Clinical Atherosclerosis Research, Department of Medicine, University of Innsbruck, A-6020 Innsbruck, Austria.

The clinical relevance of apolipoprotein determination

H. Wieland

Medizinische Universitätsklinik, Zentrallabor, Freiburg, Federal Republic of Germany

Summary

At the time being the only clinical relevance of the determination of apolipoproteins appears to be the use in monitoring the course of a disease of lipid metabolism. Apolipoproteins AI and B are very stable parameters which can be determined with high precision. If one could agree on one method for determination of apo B or establish a central laboratory, the determination of apo B in dried blood-spots for screening purposes is desirable. By this way not only cases of familial hypercholesterolemia could be detected very early in the infant, but also early enough in one of the parents. This is meaningful since we have adequate means of treating this disorder.

Lack of standardization and of agreement on methodology so far has hindered prospective data regarding the clinical significance of apolipoprotein measurement.

Keywords: Apolipoproteins, clinical relevance, chances, limits.

Introduction

The quantification of large proteins without enzymatic activity in body fluids by immunological means is a complicated matter. Whereas the quantification of peptides or synthetic antigens by RIA or ELISA techniques has found widespread use, only ferritin, transferrin or immunoglobulins are routinely quantified as accurately and as precisely as possible.

Whereas it is not difficult to distinguish between a gross abnormal concentration and normal concentrations (CRP, α_1-antitrypsin), exact quantification of serum proteins to yield a value below or above a certain cut-off point, poses several problems.

Different methods, different antibodies and different standards act together to complicate the matter. This is especially true for the immunological quantification of apolipoproteins. Standardization is very difficult, since it is almost impossible to use a "native" standard, i.e. a protein, which is in the same physicochemical state as the antigen to be assayed. The lack of adequate standardization procedures and of agreement upon which method should be used, has led to the fact, that up to now no reference range for the concentration of different apolipoproteins has been established.

Therefore, we have no prospective data concerning the correlation of apolipoprotein concentrations with disease. This is a pity, since we now have the technical means to incorporate the determination of apolipoproteins in routine clinical chemistry. The candidate apolipoproteins are apo B and AI, the major proteins of low density lipoproteins (LDL) and high density lipoproteins (HDL) respectively.

Determination of apo B and apo AI

Since it is most likely that each apo B-containing lipoprotein particle contains only one molecule of apo B per particle, the concentration of apo B reflects the amount of potentially atherogenic lipoprotein particles. The methods for detection of apo B do not need to be very sensitive, since the concentration is usually around 100 mg/dl. Thus, many different methods would be suitable. For automation on a random access analyzer immunoturbidimetry or nephelometry seem to be the methods of choice. The problem, however, is that apo B can be part of particles differing tremendously in size, so the cross-linking by antibodies may even reduce the turbidity of the sample. Therefore, determination of apo B is most useful in normotriglyceridemic samples.

Whereas apo B correlates well with the concentration of the major cholesterol carriers in the plasma (LDL), and could therefore replace the cholesterol determination if one wants to obtain a rough estimate of low density lipoprotein cholesterol (LDL-C), apolipoprotein AI correlates very well with high density lipoprotein cholesterol (HDL-C).

The concentration of apo AI is even higher than that of apo B and in addition, apo AI-carrying lipoproteins have a rather uniform size. This gives rise to the hope that standardization might be successful within the next two years. Up till now it has not been possible to determine HDL-C by a homogeneous assay as would be required by random access analyzers. Thus, the determination of apo AI would be the ideal substitute for HDL-C. Unfortunately, we do not have any prospective data.

Diagnosis, management and screening

Clinical chemistry results are used for the three major purposes diagnosis, management and screening.

Diagnosis

Since a diagnosis, which heavily depends on laboratory results, usually requires a very accurate and precise method, at the time being, not many diagnoses can be obtained by apolipoprotein determination. Of course, the absence or near absence of an apolipoprotein can be easily detected (hypoalphalipoproteinemia, Tangier disease, abetalipoproteinemia, hypolipoproteinemia). The same holds true for an excessive concentration (familial hypercholesterolemia).

Management

For monitoring of a hypolipidemic therapy, only high precision is required. Most clinical chemical tests in large hospitals are performed not for diagnostic, but for management purposes. If physicians could divert their attention from cholesterol to apo B-containing lipoproteins, the success of any lipid-lowering therapy could easily be assessed by the determination of apo B and apo AI. This even holds true for hypertriglyceridemia. At the time being, this could be the major purpose of incorporating apolipoproteins in clinical chemistry. The majority of cases of premature coronary heart disease is not due to gross hyperlipidemia, but to a common condition which can be called atherogenic dyslipoproteinemia. This eventually life-threatening disease consists of a single feature or of a combination of abnormalities. These are:

1. too much LDL,
2. not enough HDL,
3. atherogenic LDL particles,
4. not enough antiatherogenic HDL particles,
5. too many remnants of triglyceride-rich lipoproteins or
6. too much Lp(a).

Conditions 3 to 5 can only be identified by determination of apolipoproteins in addition to lipids and lipoproteins. The first two conditions could be also established by only determining apolipoproteins, provided a standardized method exists.

Screening

Because of the major importance of atherogenic dyslipoproteinemia, German Health Authorities are now supporting screening for cholesterol and HDL-C. This could much more elegantly be done by using apo B and apo AI. The only way to detect familial hypercholesterolemia (FH) in blood spots of filter paper collected for screening of other inborn metabolic disorders is the determination of apo B by a very sensitive immunoassay. This could also be applied to cord-blood. Since newborn screening is compulsory, one could detect a large majority of young adults with FH by means of their children. Detection would usually come at an early age. Since adequate therapy of FH is now available, coronary heart disease in these individuals could be postponed to a "normal" age. In case of newborn screening, one can expect a bivariate distribution of apo B. Therefore, the cut-off point can easily be identified. The determinations of apo B could be centralized nationwide. Therefore, the problems of standardization seem to be negligible.

Characteristic features of atherogenic dyslipoproteinemia

To detect the other features of atherogenic dyslipoproteinemia, it is necessary to investigate the apolipoprotein content of serum fractions. Atherogenic LDL may have abnormalities in the lipid content, which can be detected by the ratio of apo B to the several lipids in the LDL fraction. The fact that many lipoprotein particles do not contain a single apolipoprotein can be exploited by using sandwich ELISAs or determination of apolipoproteins in fractions, which have been freed of other apolipoproteins by immunabsorption. Using these procedures, the amount of HDL containing apo AI, but not apo AII could be established, these particles being more anti-atherogenic than those containing both apolipoproteins. A characteristic feature of remnants is the possession of apo B and apo E. Quantification of these particles may substantially improve risk assessment.

Lipoprotein(a) seems to be the single most atherogenic lipoprotein. Its concentration does not change very much. Therefore, exact quantification of Lp(a) at young age seems to be desirable. It appears, however, that the Lp(a) particle by itself correlates with the atherogeneity rather than the concentration of its apo(a) content. The only constant apolipoprotein component is apo B. Therefore, apo B should be used as a second antibody in the quantification of Lp(a) by a sandwich ELISA.

A characteristic feature of chylomicron remnants is the presence of apo B-48. This apolipoprotein roughly represents the N-terminal half of apo B-100, the major apolipoprotein of VLDL and LDL particles. By quantification of total apo B and apo B-100 through specific monoclonal or polyclonal antibodies, the amount of remnants may be estimated using not fractions or particles, but using whole serum. This possibility in combination with an oral fat load may help identify individuals suffering from a delayed remnant clearance and therefore an increased risk for CHD.

Conclusion

These considerations show that at the time being the clinical relevance of the determination of apolipoproteins predominantly lies in newborn screening and monitoring of therapy, since standardization does not seem to be crucial. Once exact standardization at least for the determination of apo B, apo E, apo AI and apo A-II has been achieved, the clinical relevance will increase considerably, especially in the diagnostic area.

Author's address: Dr. H. Wieland, Medizinische Universitätsklinik, Zentrallabor, Hugstetter Straße 55, D-W-7800 Freiburg, Federal Republic of Germany.

Development of reagents for apolipoproteins AI and B designed for automated routine determination

J. Karl

Boehringer Mannheim GmbH, Mannheim, Federal Republic of Germany

Summary

Two improved tests for the automated determination of apolipopropteins AI and B on BM/Hitachi analyzers were developed. Both assays are based on the immunoturbidimetric technique. Several typical properties of turbidimetry and the resulting potential for optimization are discussed. The development of a special enhancer detergent mix leads to rapid, reproducible assays with a two-week calibration stability and excellent performance data.

Keywords: Reagents for apo AI and B, turbidimetry, automated routine determination, high-dose hook effect, reaction enhancement.

Introduction

In 1987 Boehringer Mannheim launched assays for the determination of apolipoprotein AI (apo AI) and apolipoprotein B (apo B) within their CBR program.

The growing interest in apolipoproteins and the requirements for measuring apolipoproteins in clinical routine laboratories stimulated us to develop new rapid assays for the BM/Hitachi analyzers as well as other analyzers.

Why turbidimetry?

Apo AI and B are proteins which are found in high concentrations in serum. Their molecular weight allows them to be determined by classical agglutination immunoassays.

The principle of this basic reaction, called the immunoprecipitation reaction, was investigated by Heidelberger and Kendall [1] more than 50 years ago. Analyte and antibody form a three-dimensional, insoluble immunocomplex, which can be measured by turbidimetry or nephelometry (Fig. 1).

In the case of turbidimetric detection, the decrease in incident light transmitted through a suspension of particle is measured.

Immunoturbidimetric measurement has expanded rapidly in the recent years, due to the following advantages:

- High specificity due to the use of specific antibodies
- Adaptable to practically any photometer
- Automation is possible on most clinical chemistry analyzers
- Liquid, ready-to-use reagents can be developed
- Stable calibration curve
- Adequate sensitivity for a large number of plasma proteins

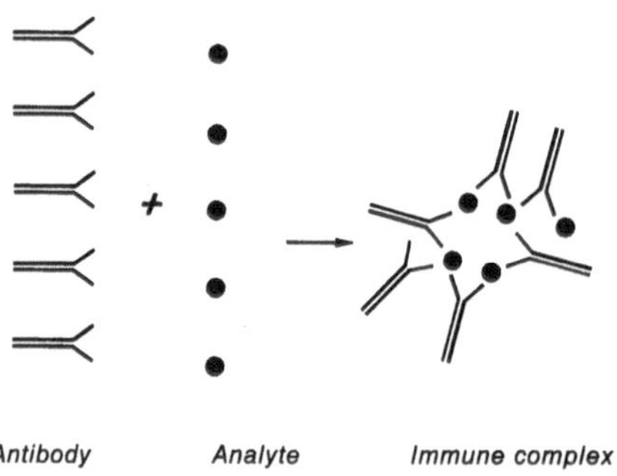

Fig. 1. The immunoprecipitin reaction

These are arguments to select the turbidimetric technique for the determination of apo AI and B.

Typical properties of turbidimetry and potential for optimization

Preciptin curve

The reaction of antibody and antigen is described by a typical precipitin curve according to Heidelberger and Kendall (Fig. 2). By increasing the analyte concentration and maintaining a constant amount of antibody, the point of equivalence is passed. This is the point, at which all antibodies are bound to the antigen binding sites.

The left side of the precipitin curve illustrates the region of antibody excess. Normally, the calibration curve represents a part of this region. Because of the typical shape of the precipitin curve, the calibration curve cannot be linear and

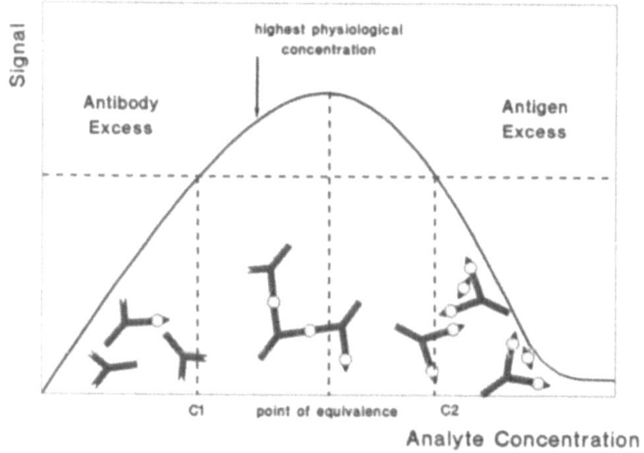

Fig. 2. Typical precipitin-curve and high-dose-Hook-effect in turbidimetric determinations

turbidimetric assays must be calibrated by more than three points.

High-Dose Hook effect

From Fig. 2 it becomes clear that there are potentially two analyte concentrations for each signal, one on the antibody and one on the antigen excess side of the precipitin curve. This means that a very high analyte concentration will be measured erroneously as a low concentration. In immunology, this effect is described as the High-Dose Hook effect. This phenomenon can be prevented by optimizing the reaction conditions.

In order to achieve this in our assay, the point of equivalence is placed above the highest physiological concentration of the analyte. Therefore, no antigen excess detection methods such as prozone-check or restart with sample are necessary.

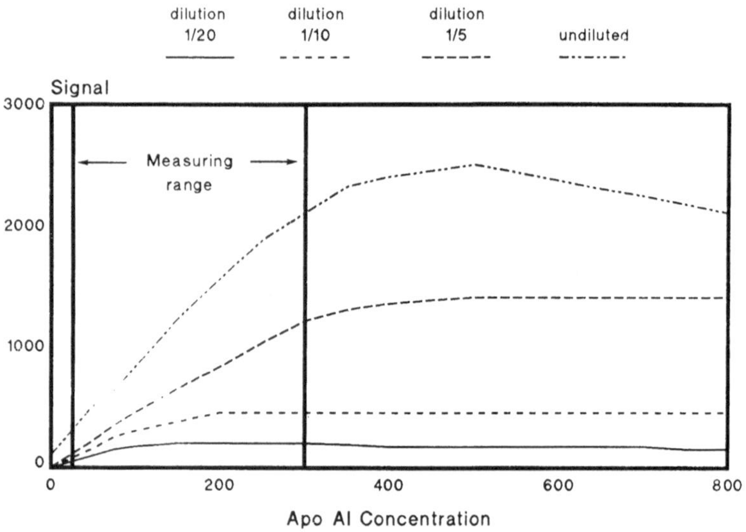

Fig. 3. The effect of antibody concentration on the precipitin curve

Optimal antigen/antibody ratio

The shape of the precipitin curve depends upon the origin of the antibody, its avidity and its concentration. Horse antisera, for example, have narrower peaks of equivalence than sheep antisera [4].

Particularly, the titer of specific antibody or the antibody/antigen ratio has an important influence on the precipitin curve. With increasing antibody concentration three effects can be observed (Fig. 3):

– an increase in the signal and therefore increased sensitivity;

– a shift of the point of equivalence to higher analyte concentration;

– a linearization of the calibration curve.

However, in most cases the serum concentration of the analytes is so high that an increase in antibody concentration is not possible because:

– the increase of antibody concentration leads to a dramatic increase in production costs (nearly 60% of the production costs are accounted for by the antibody);

– for a company with high sales rates it is impossible to produce large enough amounts of high quality antibody with respect to specificity, titer and avidity.

It is thus more economical to reach an optimal antigen/antibody ratio by reducing the analyte concentration. This is done by sample predilution.

The following arguments support this step:

- Decreased production costs and therefore lower test price
- Less variation in antibody lot
- Reduction of interference caused by particles in the sample
- Possibility of increasing the measuring range
- Avoidance of High-Dose Hook effect

– Linearization of the calibration curve (1-point calibration)

– Sample predilution can mainly be carried out by the analyzer

Enhancement of antibody-antigen reaction

The enhancing influence of different polymers such as PEG, dextran, polyvinylalcohol, etc. on the precipitin curve was described by Hellsing [2]. The mechanism favoured by Hellsing as enhancing immunoprecipitation was the steric exclusion concept.

There are two factors involved in this effect:

– Protein molecules become less soluble, an effect that is directly related to the molecular size.

– The polymer molecules strongly increase the local con-

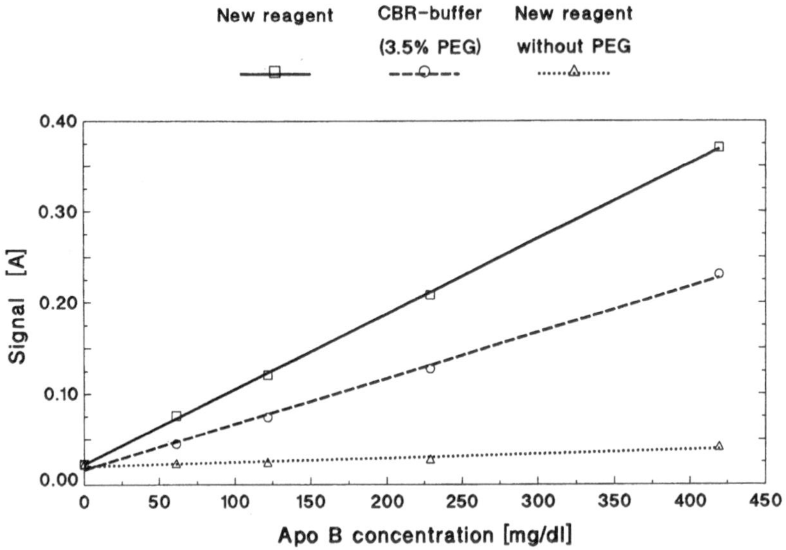

Fig. 4. The influence of the buffer composition on the calibration curve of apo B

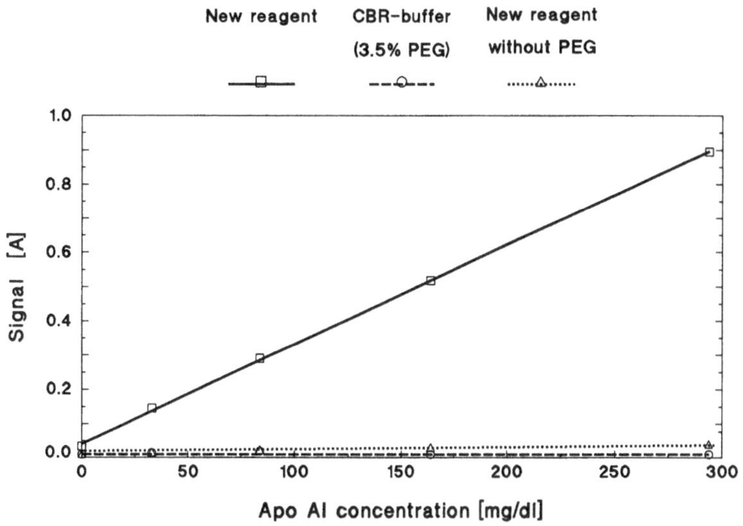

Fig. 5. The influence of the buffer composition on the calibration curve of apo AI

centration of antigen and antibody by occupying a large amount of space, thus driving the reaction towards the formation of complexes.

Figure 4 illustrates the influence of buffer composition on the calibration curves of the apo B assay.

The buffer without enhancer shows little turbidity. A strong increase in the turbidity is seen with the use of the CBR-buffer, containing 3.5% PEG 6000 in a special detergent mix.

A further increase can be achieved with new buffer, which produces a more sensitive assay with better precision in comparison to the CBR assay. This is the result of an acceleration of the antibody antigen reaction, an effect which is important due to the short incubation time on the BM/Hitachi analyzers.

The endpoint is almost reached within 5 minutes. The effect of polymer enhancement is also influenced by the type of antigen. In contrast to apo B, the effect of different buffers in the apo AI assay is more dramatic (Fig. 5).

The buffer without PEG and the CBR buffer with 3.5% PEG 6000 produce a minimal reaction within this short incubation time. The reason for this is not known.

The great difference in the molecular weights of apo AI and B, 28,500 and 270,000 respectively may, however, contribute to this phenomenon. With our new reagent containing a special enhancer and detergent mix, an enormous improvement can be achieved. This reagent also prevents interference by lipemic samples.

Antibody requirements

The highly sensitive system described above requires an antibody solution of high quality. Special concentration and purification steps are necessary to produce material of an adequate quality. Because of their high avidity, we selected polyclonal sheep antibodies. The use of monoclonal antibodies in turbidimetric assays is only possible with a mixture of more than three different ones [3]. Therefore, monoclonals are not economical for use in turbidimetric assays.

It is typical for an immunological assay that the calibrator values depend on the antibody lot. Therefore, we routinely determine calibrator values for each existing antibody lot. By packaging the calibrators with the kit, the possibility of a mix up can be excluded and high accuracy attained.

Conclusion

We have developed a rapid, reproducible assay with ready-to-use reagents and two-week calibration stability.

No dilution of buffers or antiserum with occasional filtration is necessary. Applications are available for many analyzers such as BM/Hitachi 704, 705, 717, COBAS®, Technicon RA 1000 and Eppendorf Epos.

References

1. Heidelberger M, Kendall F (1935) Quantitative theory of the precipitin reaction: study of azoprotein – antibody system. J Exp Med 62: 467
2. Helbing K (1974) The effect of different polymers for the enhancement of the antigen-antibody reaction as measured with nephelometry. Protides Biol Fluids Proc Collog Vol 21: 579
3. Pruvot I, Fievet C, Durieux C, Vu Dac N, Fruchart J-C (1988) Electroimmuno- and immunonephelometric assays of apolipoprotein AI by using a mixture of monoclonal antibodies. Clin Chem 34/10: 2048
4. Whicher J, Price C, Spencer K (1982) Immunoephelometric and immunoturbidimetric assays for proteins. CRC Crit Rev Clin Lab Sci 18/ 3: 213

Author's address: Dr. J. Karl, Boehringer Mannheim GmbH, Sandhofer Straße 116, D-W-6800 Mannheim 31, Federal Republic of Germany.

Results from a European multicentre evaluation of immunoturbidimetric assays for apolipoproteins AI and B

J. Jarausch [1], E. Casals [2], D. Gnat [3], H. Drexel [4], F. X. Huchet [5], J. Patsch [4], and H. Wieland [6]

[1] Boehringer Mannheim GmbH, Evaluation Department, Mannheim, Federal Republic of Germany; [2] Villaroel, Barcelona, Spain; [3] Hôpital St. Pierre, Brussels, Belgium; [4] Medizinische Universitätsklinik, Innsbruck, Austria; [5] Centre des examines de santé, Poitiers, France; [6] Medizinische Universitätsklinik, Freiburg, Federal Republic of Germany

Summary

Two immunoassays for apolipoproteins AI and B were evaluated in 5 European laboratories. The assays are based on the immunoturbidimetric detection of the precipitin reaction between apolipoproteins and specific anti-apolipoprotein antibodies at 340 nm. The measurements with the new reagents were performed on BM/Hitachi 704/717 and COBAS® Mira analysers. In method comparison with automated routine assays, a reasonable comparability was obtained with apo AI, whereas larger discrepancies were found with some of the routine procedures for apo B. The results indicate that the new reagents can be successfully integrated into standard lipid diagnostics performed on clinical chemistry analysers.

Keywords: Apolipoproteins, method comparison, interference, multicentre evaluation.

Introduction

Diagnostic tools for the assessment of disorders in lipid metabolism are frequently applied in the daily routine in clinical chemistry laboratories. The determination of serum cholesterol and triglycerides for a first look at the lipid metabolic state can be performed by simple procedures of high analytical reliability, i.e. reproducibility and accuracy. The reproducibility is generally poorer with immunological tests, such as the determination of protein components of lipoproteins. These determinations are applied for a more distinct look at the metabolic situation [1, 7, 9, 10]. Automation has improved the reliability of immunological procedures and the comparability of results obtained at different laboratory sites. The prime task of the European multicentre evaluation was to investigate the analytical performance of two new immunological reagents for the determination of apolipoproteins AI and B. These reagents were specially designed for automated systems used in routine clinical chemistry determinations [4].

Program of the multicentre study

The evaluation program was investigated at five European evaluation sites. Each laboratory evaluated reproducibility, analytical range limits, recovery in control samples and method comparisons with routine procedures, and participated in an interlaboratory survey. Additional tasks were investigation of the stability of calibration, comparison of 2-point – versus 5-point-calibration, stability of samples and interference studies. These tasks were only evaluated by some of the evaluators and most of the interference studies were done in the evaluation department of Boehringer Mannheim.

Material and methods

The new immunological assays for the apolipoproteins AI and B (Boehringer Mannheim, Cat. Nos. 1174371 and 1174380 respectively) are designed for endpoint determinations on clinical chemistry analysers [4]. The turbidity caused by precipitates from apolipoproteins and specific antibodies is measured at 340 nm. Tensides are included in the formulation of the reagents to improve the accessibility of antigens. The determinations can also be performed manually. Before incubation with antiserum, the samples have to be diluted 21 fold by isotonic sodium chloride solution.

Primary standardization was performed using IUIS reference material provided by CDC (CDC 1883). Calibration standards as well as control samples (Precinorm® L, Cat. No. 781827) were based on lyophilized human serum pools. The standard material is related to a single production lot of antiserum and therefore integrated into the reagent kit. The assay of apolipoprotein AI was calibrated via a 5-point-calibration curve (four parametric logit/log) and the method of apolipoprotein B was calibrated in two ways, either via 5-point- or via 2-point-calibration.

Comparison methods were assays established in routine use in the laboratories. These routine assays were based on turbidimetric (laboratory 2), rate immunonephelometric (laboratories 1 and 4) and another immunonephelometric detecting principle (fixed-time kinetics; laboratories 3 and 5). Each of these commercial assays was calibrated using the dedicated standard material offered by the manufacturers. Frozen and fresh serum samples were used. Dilution of serum samples was performed immediately before the measurements.

Results

Reproducibility of calibration curves

The reproducibility of calibration was investigated in order to estimate the frequency of calibration runs suitable for clinical routine application. Over a period of 30 days two control samples were determined in 22 runs (Fig. 1). Each sample was measured twice, either related to a calibration performed with each run or related to a (fixed) calibration determined on the first day. As shown in the figures there is no significant difference in terms of single or frequent calibration. The recoveries were reproducible over the whole period of 30 days. From these results it is suggested that a single calibration procedure is valid for at least 2 weeks.

Reproducibility

Three human serum pools at low, intermediate and high levels of apolipoproteins as well as the lowest standards were investigated. Inter-assay imprecision was determined using frozen aliquots of these pools. Effects caused by rethawing may

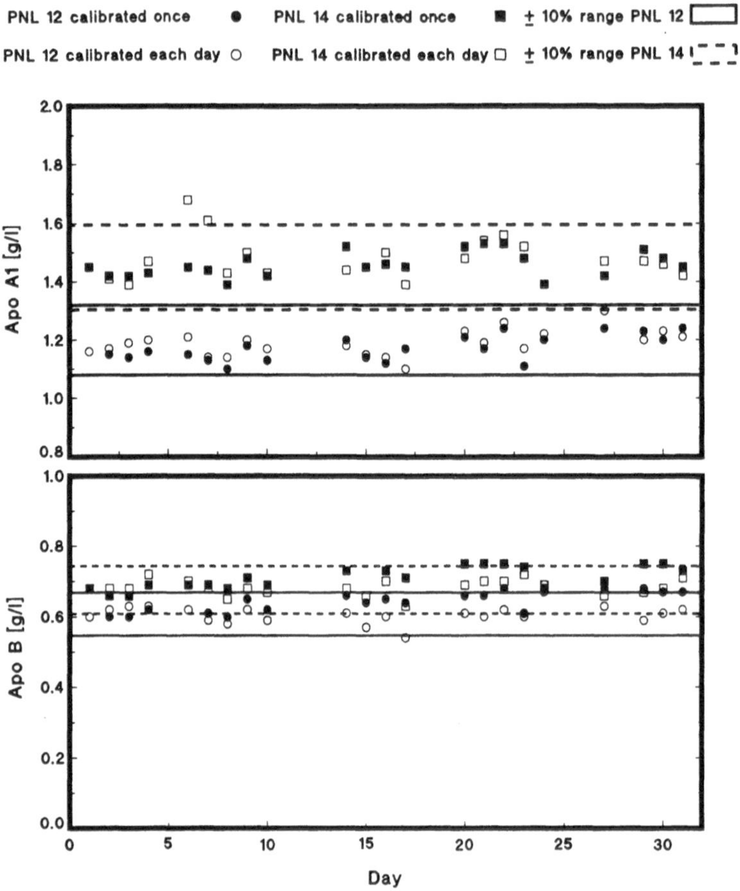

Fig. 1. Reproducibility of calibration of the assays for apo AI and apo B shown in the recovery in control sera. Control sera were measured twice, the reagents being calibrated either each run or only on the 1st day

contribute to the coefficients of variation presented in Table 1. With both methods the imprecision was close to that of standard substrate determinations in clinical chemistry. With the reagent for apo AI, the imprecision was slightly better than with the reagent for apo B, with maximum coefficients of variation between-run of < 5% and < 7%, respectively. The

Table 1. Reproducibility of the new apolipoprotein assays. CVs were calculated from 21 determinations (intra-assay imprecision) or the second of 2 measurements in 10 independent runs calibrated separately (inter-assay)

Sample	Concentration	CV Apo AI [%]	CV Apo B [%]
Intra-assay imprecision			
Serum pools			
	<0.5 g/l	2.2–2.7	1.8–11.6
	0.5 to 1.0 g/l	1.8–2.0	1.4–3.0
	>2.0 g/l	1.0–1.4	0.9–3.6
Lowest standard			
	approx. 0.4 g/l	3.4–4.2	2.4–3.5
Inter-assay imprecision			
Serum pools			
	<0.5 g/l	–	5.7–6.3
	0.5 to 1.0 g/l	0.9–2.4	1.9–4.3
	>2.0 g/l	1.6–3.4	2.6–4.7
Lowest standard			
	approx. 0.4 g/l	1.5–4.1	2.0–6.7

median inter-assay variation of all laboratories was below 5% in both cases. The imprecision of the new turbidimetric assays is comparable to that of the investigated immunonephelometric procedures as measured on a clinical chemistry analyser providing automated predilution of samples (inter-assay CVs < 3%).

Analytical range limits, linearity

Linear dilution series of human serum samples and isotonic sodium chloride were measured. Linear regression curves were calculated from the values of each dilution step, and the deviation of individual values from these curves were used to estimate the lower and the upper linearity limits (decision

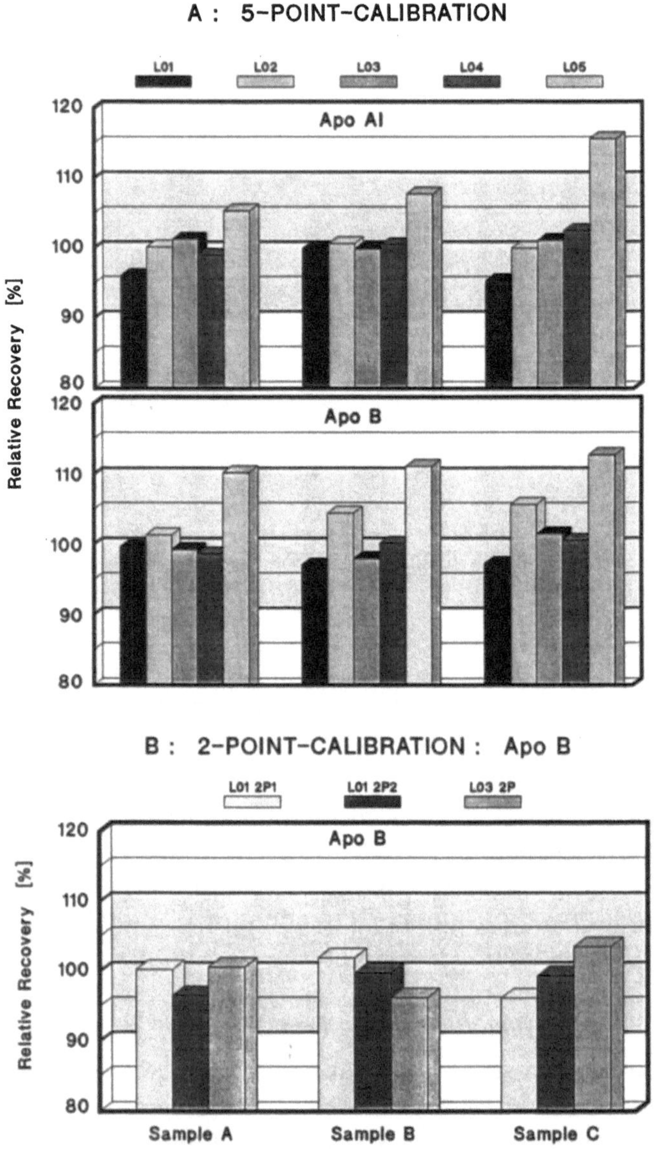

Fig. 2. Interlaboratory survey of three samples of unknown concentration

range: deviation of ± 10% from the regression curve). The limits for apo AI were 0.3 to 3.2 g/l and for Apo B (5- and 2-point-calibration) 0.3 to 4 g/l.

Accuracy

Recovery in controls

Two control samples were determined in each series of measurements. In all analytical runs the recoveries of target values were within a range of ± 20% of the target values. The median recoveries corresponded well with all instruments participating in this study.

Interlaboratory survey

Lyophilized serum samples of unknown concentration were determined in four laboratories in five independent runs. The samples of three different concentration ranges were reconstituted immediately before each run to avoid effects caused by sample storage. From the median recovery of all laboratories, the relative deviation of each laboratory was calculated. The results are presented in Fig. 2. The results obtained at different laboratory sites corresponded well, with the exception of laboratory 5, which was erroneously working with wrong calibration values (the results were recalculated for the right calibration values).

Method comparison with routine procedures

In comparing methods for the determination of apolipoproteins only relatively small discrepancies were found. On the other hand, the comparability between some comparison methods was poor [10]. This was more pronounced in the case of apolipoprotein B. Different primary standardization contributes to this situation [2, 11]. Both larger and smaller differences were found in method comparison studies in this multicentre evaluation, reflecting the current situation. The

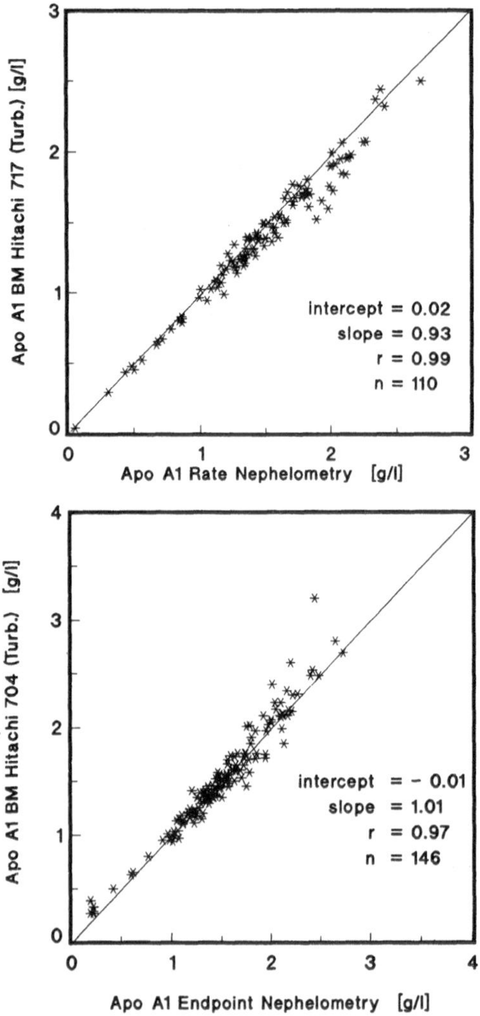

Fig. 3. Method comparison of the new assay for apo AI with two nephelometric assays

new apolipoprotein methods were standardized using IUIS reference material provided by the Centers for Disease Control (CDC, pool 1883, ref. 10). As one might expect, the

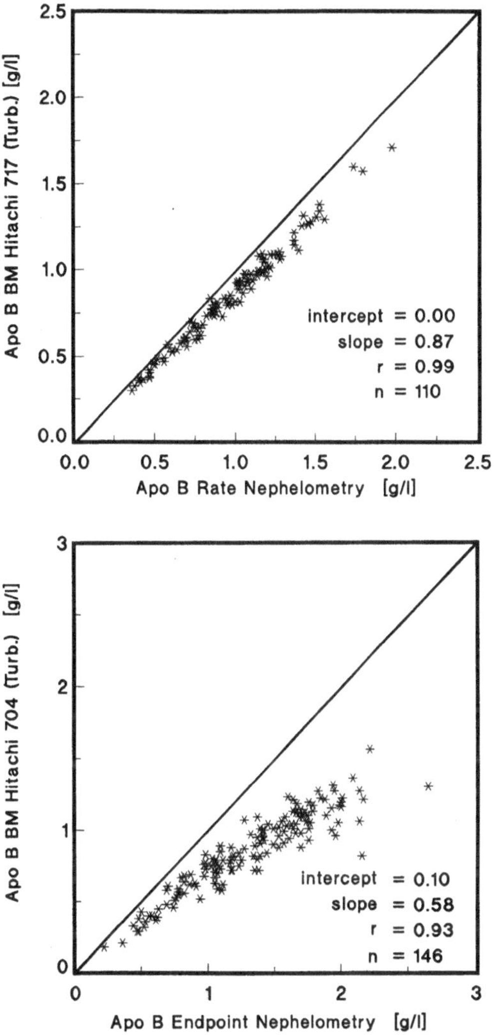

Fig. 4. Method comparison of the new assay for apo B with two nephelometric assays

comparability of apolipoprotein values measured by the new reagents and routine methods also based on CDC reference material was better (2 of the investigated procedures) than

Table 2. Method comparison of the new turbidimetric assays with nephelometric or turbidimetric assays range: [g/l]

Method Comparison Apo A1 / B (Routine Methods vs BM)								
		APO A1						
Methods	n	range min	max	intercept	slope	r	median of rel. differences	Sy.x
X = Rate Nephelometry Y = BM Hitachi 717 (Turb.)	110	0.05	2.68	0.02	0.93	0.99	-5.4	0.056
X = Technicon RA 1000 (Turb.) Y = BM Hitachi 717 (Turb.)	126	0.38	2.55	0.03	1.11	0.96	13.3	0.101
X = Fixed Time Nephelometry Y = BM Hitachi 704 (Turb.)	146	0.19	3.2	-0.01	1.01	0.97	0.5	0.085
X = Rate Nephelometry Y = COBAS (R) Mira (Turb.)	175	0.40	2.55	0.18	0.89	0.99	1.70	0.035
X = Fixed Time Nephelometry Y = BM Hitachi 717 (Turb.)	51	0.03	2.37	-0.08	1.25	0.98	19.4	0.072
		APO B						
X = Rate Nephelometry Y = BM Hitachi 717 (Turb.)	110	0.3	1.98	0.00	0.87	0.99	-13.4	0.027
X = Technicon RA 1000 (Turb.) Y = BM Hitachi 717 (Turb.)	126	0.29	3.48	0.03	0.89	0.95	-6.9	0.091
X = Fixed Time Nephelometry Y = BM Hitachi 704 (Turb.)	146	0.18	2.65	0.10	0.58	0.93	-33.0	0.074
X = Rate Nephelometry Y = COBAS (R) Mira (Turb.)	175	0.25	1.85	0.01	0.93	0.99	-6.2	0.022
X = Fixed Time Nephelometry Y = BM Hitachi 717 (Turb.)	51	0.17	4.52	0.11	0.67	0.79	-23.0	0.141

in the case of the other methods (Table 2). The latter were based on primary standards related to a separate reference survey. The correlation between the investigated methods for apo AI was generally high ($0.96 \leqslant r \leqslant 0.99$). This was not the case in comparisons with methods for apo B ($0.79 \leqslant r \leqslant 0.99$). Beside different primary standardization, there are also methodological influences contributing to reduced comparability of apolipoprotein values determined by different assays [2, 8]. This might also be suggested from the method comparison plots presented in Fig. 3 and 4. The deviation of individual measurements from the regression curves was relatively small in the comparison of the new turbidimetric methods with the kinetic nephelometric assays. This is not the case in the other method comparison plots. Despite the reported differences between certain methods, the results nevertheless demonstrate that a high degree of comparability of apolipoprotein values is achievable even when different detecting principles are applied.

Comparison of applications

The comparability of applications was investigated at two evaluation sites. The impact of different calibration protocols using either 5-point-calibration curves or linear 2-point-calibration is shown for applications of the new assay for apolipoprotein B to BM/Hitachi 704 (Fig. 5). As demonstrated in the figure, the results corresponded quite well. Hence, a linear calibration protocol, which is more convenient for routine use, may be used with apo B. The assay for apolipoprotein AI has to be calibrated via a 5-point calibration curvc, duc to non-linear dose/response relationship (results not shown). The comparison of applications to Hitachi 717 with the manual procedure is presented in Fig. 6. Manually determined apolipoprotein values were in good agreement with corresponding values determined by the automated procedures.

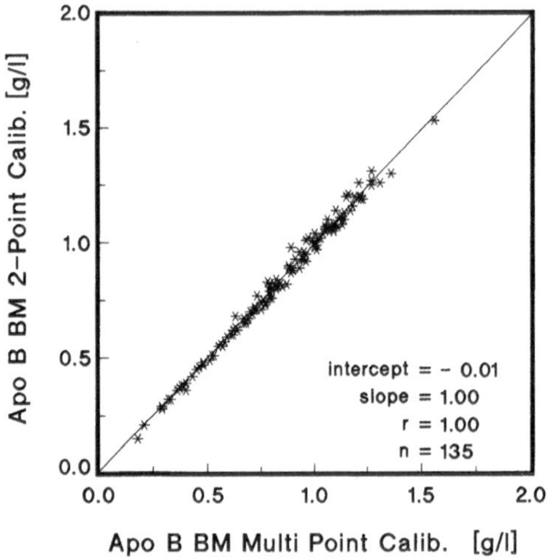

Fig. 5. Comparison of two applications of apo B to Hitachi 704 calibrated either by 2-point or by 5-point-calibration (Lab 03)

The impact of the measuring temperature was investigated on the BM/Hitachi 717. By regression analysis no significant deviation was found in apolipoprotein values obtained with the same instrument settings at different temperatures (i.e. 25/30/37 °C; results not shown).

Interference

"Classical" sources of interference were investigated by the evaluation department of Boehringer Mannheim. The influence of bilirubin, haemolysis and lipaemia was simulated in accordance with the procedure of Glick et al. [3]. In these experiments we measured linear dilution curves of serum and serum spiked by the potentially interfering substance. No interference was detected on BM/Hitachi analysers (bichro-

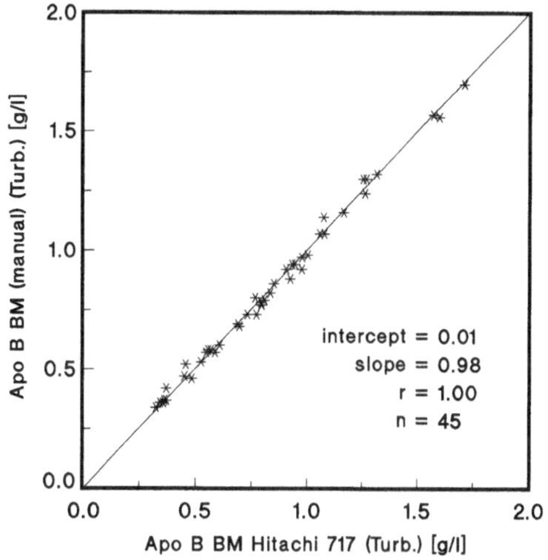

Fig. 6. Comparison of automated determination of apo B on Hitachi 717 with the manual determination

matic detection), whereas monochromatic detection of the precipitin reaction with apolipoprotein B was interferred with high levels of lipaemia. In our experiment using Intralipid®, interference was found at triglyceride concentrations above 1500 mg/l.

The interference was negligible in the case of the monochromatic determination of apo AI. It has been reported that the size of lipoproteins in hypertriglyceridemic samples affects the detection of apolipoproteins by radial immunodiffusion and nephelometry [1, 5, 6]. From our measurements we conclude that this source of analytical error is eliminated in the measurements with the new turbidimetric methods on bichromatic (BM/Hitachi) analysers. In investigating 34 drugs at elevated levels (slightly above the therapeutic concentration range) no effect on the immunoprecipitin reaction with apo AI and B was found.

Comparison of apolipoprotein values in serum and plasma

Apolipoprotein values from 30 serum samples were compared with corresponding plasma values. In this study blood was consecutively collected into standard sample carriers either uncoated or coated with anticoagulants. Values from Li-heparinate and EDTA plasma deviated slightly from the corresponding serum values (mean difference < 9%). In contrast, the values in citrate plasma deviated significantly (> 20%).

Stability of samples

The effect of storage conditions on the determination of apolipoproteins was investigated in five serum samples. Serum samples were distributed into three series of sample vials. These series were kept either at 25 °C, 4 °C or frozen at − 20°C for a period of 24 days. With frozen samples, rethawing was avoided. In samples stored at 25 °C a slight increase of about 5 to 10% in the apolipoprotein values was found for apo AI and B. A drastic decline in apolipoprotein values was found in one sample, starting from the 12th day. This was presumably caused by bacterial contamination.

A small decline of about 5 to 10% related to the fresh sample was found in the values of both apo AI and B when sample were stored either at 4 °C or at − 20 °C. From these results it is assumed, that serum samples may be stored for at least 2 weeks at 4 °C without any significant effect on apolipoprotein values. In frozen samples a slight decrease in values of apo AI and B may occur in the course of longer storage periods. In this case it is recommended to store samples at − 70 °C [12]. The sample material should not be stored at room temperature for longer than 5 days.

Conclusion

Two new turbidimetric assays for the determination of apolipoproteins AI and B have been evaluated in five European laboratories. Reproducibility and accuracy of the new assays as evaluated on different analyseres were close to standard substrate determinations in clinical chemistry. Analytical reliability was comparable to that of automated nephelometric procedures established at the evaluation sites for the determination of apolipoproteins. The elimination of lipid interference in the new turbidimetric assays improves the clinical value of the determination of apolipoproteins. From the results generated at the different evaluation sites it is concluded that the new assays can be integrated into the routine workload on clinical chemistry analysers. This drastically reduces investigational work, for example in the screening of patients for coronary artery disease risk.

References

1. Adolphson JL, Albers JJ (1989) Comparison of two commercial nephelometric methods for apoprotein AI and apoprotein B with standardized apoprotein AI and B radioimmunoassays. J Lipid Res 30: 597–606
2. Bachorik Ps, Kwiterovich Jr PO (1989) Apolipoprotein measurements in clinical biochemistry and their utility vis-a-vis conventional assays. Clin Chim Acta 178: 1–34
3. Glick MR, Ryder KW, Jackson SA (1986) Graphical comparisons of interferences in clinical chemistry instrumentation. Clin Chem 32(3): 470–475
4. Karl J, Kerscher L, Engel WD, Ziegenhorn J (1989) Immunoturbidimetric determination of apolipoprotein AI and B on Hitachi analyzers. (Abstr.) Clin Chem 35/6: 1067
5. Labeur C, Shepherd J, Rosseneu M (1990) Immunological assays of apolipoproteins in plasma: methods and instrumentation. Clin Chem 36(4): 591–597
6. Maciejko JJ, Holmes DR, Kottke BA, Zinsmeister AB, Dinh DM, Mao SJT (1983) Apolipoprotein AI as a marker of angiographically assessed coronary artery disease. N Engl J Med 309: 385 389
7. Maciejko JJ, Levinson SS, Markyvech L, Smith MP, Blevins RD (1987) New assay of apolipoproteins AI and B by rate nephelometry evaluated. Clin Chem 33(11): 2065–2069
8. Naito HK (1986) Serum apolipoprotein measurements: an improved discriminator for assessing coronary heart disease. Compr Ther 13(1): 43–52

9. Riesen WF, Mordasini R, Salzman C, Theler A, Gurtner HP (1980) Apoproteins and lipids as discriminators of severity of coronary heart disease. Atherosclerosis 37: 157–162
10. Smith SJ, Cooper GR, Henderson LO, Hannon WH, Apolipoprotein Standardization Collaborating Group (1978) An international collaborative study on standardization of apolipoproteins AI and B. Part I. Evaluation of a lyophilized candidate reference and calibration material. Clin Chem 33: 2240–2249
11. Smith SJ, Cooper GR, Henderson LO, Hannon WH, Phase V (1987) Preliminary report. Apolipoprotein standardization program IUIS-CDC-NHLBI, December 1987
12. Wang XL, Dudman NPB, Blades BL, Wicken DEL (1989) Changes in the immunoreactivity of apo a-I during storage. Clin Chim Acta 179: 285–294

Author's address: Dr. J. Jarausch, Boehringer Mannheim GmbH, Evaluation Department, Sandhofer Straße 116, D-W-6800 Mannheim, Federal Republic of Germany.

Evaluators' addresses: E. Casals, Villaroel, 170, E-08036 Barcelona; D. Gnat, Hôpital St. Pierre, Rue Haute, 322, B-1000 Brussels; H. Drexel, J. Patsch, Medizinische Universitätsklinik, Angerstraße 35, A-6020 Innsbruck; F. X. Huchet, Centre des examines de santé, 21, rue Saint Louis, F-6800 Poitiers; H. Wieland, Medizinische Universitätsklinik, Hugstetter Straße 55, D-W-7800 Freiburg.

Standardization of apolipoprotein measurements

M. Rosseneu and **C. Labeur**

Department of Clinical Chemistry, A. Z. St-Jan, Brugge, Belgium

Summary

The standardization programs for the immunological assays of apolipoproteins have temptatively addressed several of the problems linked with this type of assays. Such problems include: lack of primary standards with certified values, method-to-method variation, lack of comparability for sample treatment, etc.

International organizations including: the International Union of Immunological Societies (I.U.I.S.), the International Federation of Clinical Chemistry (IFCC), the European Community Bureau of Reference (BCR), have supported these programs. Although partial answers to some of the problems have been obtained, future collaborative efforts are still required to solve the remaining questions.

Keywords: Immunoassay, primary standard, reference material, standardization.

Introduction

Since the introduction of immunological assays for apolipoproteins, their application to the measurement of apo AI and B has been hampered by the lack of reference methods and materials. This resulted in large variations between values obtained in different laboratories. Standardization is therefore urgently required in order to compare and interprete these data.

This paper will review the results from recent standardization programs and suggest directions for further collaborative efforts.

The standardization program of the International Union of Immunological Societies (I.U.I.S.)

This program was launched on the initiative of the Standardization Committee of the I.U.I.S. in 1981 [1, 5, 9] with Co-chairmen G. Cooper (CDC Atlanta) and M. Rosseneu (Brugge). This committee functioned between 1981 and 1989 and set up five major standardization steps. The first phase of the standardization program consisted in a survey of the existing methodologies and of the published values for apo AI and B in a normal population [1, 2]. The literature survey showed that many laboratories were already measuring apolipoproteins, using a variety of immunological assays, yielding widely different values for comparable normal populations [1, 2]. The coefficients of variation reported per laboratory varied between 5 and 10% and could not account for the wide spreading of the results among laboratories.

A first survey with 55 participating laboratories was conducted in 1983 [2]. The measurement of a common lyophilized serum pool emphasized the wide interlaboratory variations when local antisera and standards were used (Table 1).

23 laboratories participated in the phase of the program for the measurement of a lyophilized serum and of fresh sera collected in each lab [3, 7]. The data collected led to the conclusion that the source of antiserum contributes only weakly to the assay variability.

The measurement of three lyophilized pools with low, normal and elevated apo AI and B levels by laboratories demonstrated that the variability increased with increasing apolipoprotein levels. In all cases the variability was higher for apo B than for apo AI, mainly due to inter-lab variation [6]. From these results it was concluded that the use of common reference material as calibrator should significantly reduce the inter-lab variability for apo AI and B measurement.

Table 1. Analysis of source of variation in the apo AI and B assays, in the I.U.I.S.-CDC standardization program

Source variation	Apo AI		Apo B	
	PCT	CV%	PCT	CV%
Replicates	39	12	17	12
Method	13	7	33	18
Laboratory	40	12	45	21
Antisera	8	5	5	7
CV total		19		31

PCT % of total variation

The serum pool prepared by CDC, however, had the major drawback of being a lyophilized material which is less suitable for apo B measurements due to matrix effects. The use of a stable liquid standard should be prefered for apo B, as developed in the frame of the International Federation of Clinical Chemistry program (IFCC).

The standardization program of the International Federation of Clinical Chemistry (IFCC)

The aim of this program was to provide a common secondary reference material to all companies manufacturing laboratory kits for the immunological measurement of apo AI and B. This secondary reference material was selected from reference materials provided by different companies and it was compared to frozen serum pools prepared by the Northwest Lipid Research Clinics in Seattle. The frozen serum pools were used to calibrate the secondary reference material so that comparable values would be obtained, independent of the technique and equipment used. The concentration values and the coefficients of variation obtained for the measurement of plasma samples were compared before and after method calibration with the frozen serum pools.

19 companies and reference laboratories participated in the first phase of the study (1989–1990), and apo AI and B were measured by immunonephelometry, immunoturbidimetry, radioimmunoassay and radial immunodiffusion. The immunoprecipitin reactions were most commonly used as the automated clinical analyzers used for routine measurements are based upon this technique. After correction of the values of the secondary reference materials to obtain uniform values for the frozen serum pools, the coefficients of variation between techniques and between laboratories decreased significantly.

A candidate reference material was selected among those proposed by the companies, which met the following criteria: measurable with a coefficient of variation less than 5%, no turbidity after reconstitution.

Preparation of primary standards for Apo AI and Apo AII. The European Community Bureau of Reference Standardization Program

In order to calibrate any frozen pool or secondary reference material, primary standards consisting of either a purified low-density lipoprotein fraction for apo B, or of purified lyophilized apoproteins for apo AI and AII [4], are required. The European Community Bureau of Reference Standardization Program was initiated in 1986 with the aim to prepare stable standards for apo AI and AII to which a certified value could be assigned. These standards should further serve to calibrate any secondary reference material.

In a pilot study (1986–1988), a small batch of purified apo AI and AII was prepared and tested. As this material was found suitable, a larger batch of 1 g apo AI and 500 mg apo AII each were made in 1989.

The certification of the mass value for the lyophilized

material was carried out by 9 European laboratories. The mass of apo AI and AII was determined by amino acid analysis and by phenylalanine quantification by HPLC. The results were approved by the Certification Committee of the Community Bureau of Reference and the certified material RM 393 and 394 contain 1.060 mg of apo AI and 0.321 mg of apo AII respectively per ampoule. These ampoules are to be reconstituted in 1 ml of phosphate buffer and used to calibrate secondary standards for apolipoprotein assays [11].

Conclusions

The standardization of apolipoprotein measurements is not an easy task, as the nature of the apolipoproteins and their association with plasma lipids adds to the difficulties encountered in similar programs developed for plasma proteins [8, 10].

The availability of the primary standards for apo AI and AII and that of secondary reference material for apo AI and B should enable the transfer of the exact value from primary standards to the secondary reference material, through reference methods. A further cooperation between the international institutions sponsoring these programs should lead to universally accepted standards for apolipoproteins AI and B in the near future. This step will certainly contribute greatly to a better use and application of apolipoprotein assays in research and clinical laboratories.

References

1. Bachorik PS, Kwiterovich PO (1988) Apolipoprotein measurements in clinical biochemistry and their utility vis-à-vis conventional assays. Clin Chim Acta 178: 1–34
2. Cooper GR, Smith SJ, Wiebe DA, Kuchmak M, Hammon WH (1985) International survey of apolipoprotein AI and B measurements (1983–1984). Clin Chim 31: 223–228
3. Henderson LU, Hannon WH, Smith SJ, Cooper GR (1987) An international collaborative study on standardization of apolipoproteins

AI and B. Part I. Evaluation of contributions of antisera to among-laboratory variance components. Clin Chem 33: 2250–2256
4. Labeur C, Shepherd J, Rosseneu M (1990) Immunological assays of lipoproteins in plasma: methods and instrumentation. Clin Chem 36: 591–597
5. Rosseneu M, Vercaemst R, Steinberg KK, Cooper GR (1983) Some considerations of methodology and standardization of apolipoprotein B immunoassays. Clin Chem 29: 427–433
6. Smith SJ, Cooper GR, Henderson LU, Hammon WH (1986) Phase IV preliminary report. Apolipoprotein standardization. International Collaborative Study IUIS-CDC-NHLBI
7. Smith SJ, Cooper GR, Henderson LU, Hammon WT (1987) An international collaborative study on standardization of apolipoproteins AI and B. Part I. Evaluation of a lyophilized candidate reference and calibration material. Clin Chem 33: 2240–2249
8. Sniderman AD, Silberberg J (1990) Is it time to measure apolipoprotein B? Arteriosclerosis 10: 665–667
9. Steinberg KK, Cooper GR, Rosseneu M (1983) Evaluation and standardization of apolipoprotein A-I immunoassays. Clin Chem 29: 415–426
10. Vega GL, Grundy SM (1990) Does measurement of apolipoprotein B has a place in cholesterol management? Arteriosclerosis 10: 668–671
11. Purification and certification of human apolipoprotein AI and AII Reference Materials (RM 393, RM 394). Report of the European Community Bureau of Reference (1990)

Authors' address: Dr. M. Rosseneu, Department of Clinical Chemistry, A.Z. St-Jan, B-8000 Brugge, Belgium.